Research and Practice

Essays of a researcher-practitioner

Željko Obrenović

Research and Practice

Essays of a researcher-practitioner

Željko Obrenović

ISBN 978-1542316859

This is a Leanpub book. Leanpub empowers authors and publishers with the Lean Publishing process. Lean Publishing is the act of publishing an in-progress ebook using lightweight tools and many iterations to get reader feedback, pivot until you have the right book and build traction once you do.

© 2016 - 2020 Željko Obrenović

To Jacobine, Nada and Marie.

Contents

Foreword i

I RESEARCH & PRACTICE . 1

1. The Curious Case Of "Small" Researchers-Practitioners 3
2. The Hawthorne Studies 13
3. Is Academia Guilty of Intellectual Colonization of Practice? 29
4. The Four Points of the Research Compass 33
5. Insights from the Past 47
6. The Researchers-Practitioners Manifesto 71

II DESIGN & RESEARCH . . . 75

7. Design-Based Research 77
8. Doing Design-Based Research in Practice 89
9. Design As a Political Activity 97

III NEW IDEAS 109

10. Experiential Learning of Computing Concepts . 111

11. Sketchifying: Bringing Innovation into Software Development 125

To Probe Further: Selected Bibliography 145

Foreword

In my career, I have been doing research in both academic and industrial settings. This experience has provided me with an opportunity to see both positive and negative aspects of doing research in practice. With this collection, I want to share some of the lessons I learned.

The central theme of the essays is the tension between the value of doing research in practice and difficulties that such work brings. Practice is vibrant and still hugely unexplored area, and researchers-practitioners may be in unique positions to witness or make important discoveries in many areas of computing. However, there are a number of barriers and challenges that practitioners-researchers face.

Due to my background and interests, essays are limited, to some degree, to discussions related to software engineering and interaction design. However, in the essays, I borrow from many disciplines, including computer science, software engineering, human-computer interaction, interaction design, classical design, and philosophy.

Structure of the Book

This book is organized into three parts.

Part I introduces four essays that explore the relationship between the research and practice in more general terms.

The first essay "*The Curious Case Of 'Small' Researchers-Practitioners*" discusses the difficulties of doing research outside mainstream research laboratories and universities. The second essay "*Is Academia Guilty of Intellectual Colonization of Practice?*" discusses one of the common anti-patterns of research-practice collaboration. In the essay "*The Hawthorne Studies and Their Relevance to Computer Science Research*" I discuss the groundbreaking contribution of the Hawthorne studies, and why I think they are still relevant for computer-science research. The essay "*The Four Points of the Research Compass*" explores the research motivation behind our research efforts, arguing that research can be motivated by a vision, curiosity, doubt or skepticism. The essay "*Insights from the Past*" discusses the value of studying the past so that it might not be reinvented but become a source of inspiration for the present and future. The essay "*The Researchers-Practitioners Manifesto*" is a call to practitioners to start doing research in practice and to organizes themselves outside current academic structures. I believe that the only way to bridge the practice-research gap is for practitioners to start doing research and to bottom-up build necessary organizations and structures.*

Part II focuses of the relation of design practice and research. The first essay "*Design-Based Research*" discusses what why design can be a basis for doing research, focusing on what we can learn when we engage in the design of computer systems. The second essay "*Doing Design-Based Research in Practice*" presents more practical advice about how to do design-based research in practice. The essay "*Design As a Political Activity*" presents often overlooked aspect of design activities - politics.

Part III presents some of my new ideas. Doing research in practice is based on learning from experience and intensive experimentation. Consequently, two remaining essays cover these two topics. The essays *"Teaching Based on Experiential Learning Paradigm"* presents some of my ideas and attempts to use experiential learning paradigm in education. Last essay *"Bringing Innovation into Software Development"* presents my work on stimulating software designers to spend more time experimenting and considering alternative ideas before deciding to proceed with engineering.

For readers interested to further explore the topic of research and practice, I also provide a selected bibliography at the end of the book.

About the Cover

The picture is from Apollo 17 mission, 11 December 1972. On this picture Astronaut Eugene A. Cernan, commander, makes a short checkout of the Lunar Roving Vehicle during the early part of the first Apollo 17 Extravehicular Activity (EVA-1) at the Taurus-Littrow landing site. This photograph was taken by scientist-astronaut Harrison H. Schmitt, lunar module pilot. The mountain in the right background is the east end of South Massif. While astronauts Cernan and Schmitt descended in the Lunar Module (LM) "Challenger" to explore the Moon, astronaut Ronald E. Evans, command module pilot, remained with the Command and Service Modules (CSM) "America" in lunar-orbit.

I selected the cover photo because it symbolizes exploration. The picture also illustrates my approach to doing research: I build experimental systems to explore new domains. I also use the paradigm of an astronaut in space to illustrate the isolation and loneliness of researchers doing research in practice, outside established research institutions.

I RESEARCH & PRACTICE

1. The Curious Case Of "Small" Researchers-Practitioners

The following article I wrote in 2013, a year after I decided to leave my academic position and take an industrial position. It represents my initial observation on value and challenges of doing research in practice.[1]

Astronaut Ronald E. Evans outside the Apollo 17 spacecraft. Wikimedia Commons

---

[1] This chapter is based on the article Research and practice: the curious case of 'small' researchers-practitioners, Communications of the ACM 56, 9 (September 2013), 38-40.

Introduction

In recent years we have witnessed more attempts at bridging the practice-research gap in computer science [5]. ACM and IEEE Computer Society [1], for example, seem to be increasingly more open to "the voice of practice." Communications now includes the Practice section. ACM Queue promotes itself as an online magazine for practicing software engineers, *"written by engineers for engineers."* ACM interactions describes its goal as to *"lay between practice and research...making...research accessible to practitioners and making practitioners voices heard by researchers."* IEEE Software defines its mission as to build the community of leading software practitioners. The International Conference on Software Engineering (ICSE) has the Software Engineering in Practice track, and, similarly, the ACM SIGCHI conference accepts case studies intended to *"specifically reach out to the practitioner communities."*

While the research-practice symbiosis seems to be flourishing, doing research as a practitioner is still not easy. It is even more difficult if research is not conducted in big companies or in collaboration with universities. Many of us are researchers-practitioners working in relatively small companies. By researchers-practitioners, I mean practitioners with clear practical tasks in their job, but who have background or skills of a researcher, obtained, for example by getting a Ph.D. or working as a postdoctoral researcher. And I call these practitioners "small" because they usually do research independently or in small teams, and cannot associate with their work research reputation and influence of their institution or companies. In small companies, we may not have a number of things that researchers in universities or big companies take for granted, [7] such as an explicit research department, budget for conferences, freedom, or even the job description and status of a researcher. But we can bring to practice the benefits of research approach, rigor, and discipline. And we can make accessible to the research community valuable insights and unique lessons from practice.

Contributions of small practitioners-researchers, however, are not always recognized and valued. Furthermore, they face a number of challenges and obstacles that researchers in big companies or in universities do not. In this Viewpoint, I want to call attention to the value of doing "small" research in small companies, and point out some of the main obstacles that such work faces.

Recognizing the Value of "Small" Research in "Small" Companies

Researchers are, in general, good in critical thinking, analysis, and dissemination of their findings. These skills, combined with practical work, can bring to their companies and the research community several benefits. Here, I discuss two characteristics of research work I find particularly relevant for small researchers: generalization and publishing.

Generalization. Normally, the goal of practice is to create a successful product, and lessons learned in this activity are restricted to the particular solution and the people involved in it. To be acceptable as research contributions, however, these lessons need to be generalized, applicable beyond original context, and useful to others (see the chapter Design-Based Research for more details about such generalized knowledge).

Generalization is not only an abstract academic goal, but it can be valuable for practice. In my previous position, I worked in a relatively small company in a department called "best practices." The primary goal of our department (one engineer, one architect, and one researcher) was to collect, generalize, and share best software development practices related to our software products. Being a relatively small company meant we did not have the luxury to repeat errors, and our department was built with the aim of maximally leveraging the lessons learned in our projects. Our task was not to simply collect these lessons, but to generalize them and make them usable and understandable to the broader audience, within and outside our company. Applying research approaches, such as using analytic generalizations, evaluations, and connecting our findings to existing work, helps significantly. Good generalizations can also help avoiding low-level technical jargon. Consequently, our work has been valuable not only for our architects and developers but also to our sales team, who were able to use some of our analyses as arguments in discussion with demanding and critical clients. In contrast to research in big companies, small researchers are closer to the "battlefield" and can more directly contribute to the company's success.

For the research community, generalizations of practical solutions on a broader scale and across multiple projects are particularly valuable. For example, we recently published an article about security patterns of integrating authentication and personalization, generalizing security implementations in several of our projects [6]. I also see a potential value of having more smaller companies sharing their "best practices," combined with additional effort of the academic community to connect and further generalize these practices. I had an opportunity to witness the value of this approach firsthand, when I was one of the guest editors for the special issue of ACM Multimedia Systems Journal on Canonical Processes of

Media Production [4]. This special issue was not only a collection of articles, but it presented a model of media production that was based on generalization of 10 companion articles describing different media production domains (each of which presented some specific media production system or project). Contributions included several media production companies, artists, and academic researchers. The resulting model significantly benefited from interaction and generalization of issues from our industrial contributors. Our industrial contributors also benefited from connecting their work to other solutions, as they were able to get new ideas and see that their issues are shared by others and that they can learn from each other's experiences. It would be interesting to see more such attempts in other domains, where small researchers would present their initial generalizations of their domains, and a broader research community would connect these generalizations to other industrial and academic work.

Publishing Results. Publishing findings from practice has obvious benefits for the research community as it enables it to obtain deeper insights about relevant practical issues, and gets more realistic overview of the state of the practice [3]. Stolterman, for example, argued that many research projects about theoretical approaches, methods, tools, and techniques for supporting interaction designers in their practice failed because they were not guided by a sufficient understanding of the nature of practice [8].

Publishing can also significantly help a small company. One of the most important values of publishing in peer-reviewed venues is receiving knowledgeable and valuable criticism. By publishing your results, you also have to make the reasoning behind your generalized claims explicit, public, and open to critical reflection and discussion, which enables receiving feedback of experts and colleagues from different communities. Publishing results can also have positive influence on company's promotion and hiring of new employees. Small companies normally cannot sponsor huge events, but presenting a paper at a conference, combined with promotion of this event by the company, may give a company a fair share of visibility and promotion for much smaller price. Small companies may also have more difficulties attracting high-quality employees, and I received unexpected encouragement to actively participate in conferences from the Human Resources (HR) department. The HR department elaborated that such activities can help the company to demonstrate the quality of its work and its people, both to potential new clients and employees.

Main Obstacles

Doing research outside universities or big companies, even when conducted with rigor and discipline, comes with a number of challenges. Finding time and resources for research in small companies is always challenging. And practice does not always recognize the value of research contributions. It may require significant time and effort to convince relevant people in your company of the potential value of doing research. Practice also needs to understand that it is not enough to simply relabel "development" as "research," and that research cannot be done properly without individuals who are disciplined and objective enough to conduct it with scientific rigor.

Less obviously, and contrary to the recent trend of openness for "the voice of practice," a small researcher-practitioner may face even bigger barriers from the research community.

Research work is difficult and incomplete if a researcher is not a part of a community of researchers. However, for researchers-practitioners coming from smaller or less-known companies, it may be difficult to become a part of such a community. First, it may be difficult to find a venue open for contributions of the practitioners. Reviewers also may be biased toward more academic contributions and methods. When you try to submit some of your work for publication in places that seem to promote strong practice orientation, you may find many of them are not open for your contributions. For example, the Communications Practice section publishes articles "by invitation only." Similarly, ACM Queue reviews articles only from authors who have been "specifically invited to submit manuscripts." This makes it practically impossible for people outside a relatively small group of elite practitioners to even try to contribute regardless of the quality of their contribution.

Another barrier from the academic side comes from stereotypes about the research process. When working for my previous company, I tried to join the ResearchGate[2], as several of my papers have been uploaded there by other co-authors. However, when trying to register with my company email address, I received the following email message: "*We've reviewed your request and regret to inform you that we cannot approve your ResearchGate account at present. As ResearchGate is a network intended for scientific and academic exchange, we ask that you sign up with an email address affiliated with your institution (e.g., university, organization, or company) or provide us with details of your independent research (e.g., research discipline and current project).*"

My email address was affiliated with my institution (a com-

---

[2]https://www.researchgate.net/

pany), in an obvious way (my name at my company domain). However, it seems a company is considered a research organization only if it is a well-known institution, and with a separate research department (for example, Google Labs, Microsoft Research, Yahoo Research, Philips Research...). This anecdote points to a problem of researchers from smaller companies who may be discriminated in their attempts to become part of the research community, and may have difficulties passing the threshold of being considered worthy of belonging to the research community. Also the notion of a research project seems to be closer to the academic environment where researchers work for several years on the same project. In practice, there may be a long-term research thread, but research contributions do not necessarily belong to an explicit project.

Conclusion

There is a potential value for both, practice and research, if we have more active "small" researchers-practitioners. With declining numbers of research positions in academia [2] we have increasing numbers of research-capable people entering small companies. Practice is rich and still hugely unexplored area, and researchers-practitioners may be in unique positions to witness or make important discoveries in many areas of computing. However, there are a number of barriers and challenges that "small" practitioners-researchers face. Practice needs to become more aware about the value of applying research rigor and discipline, and the research community must be more open for attempts of "small" researcher-practitioners to join them as equals. Educational institutions also need to think about how to educate researchers-practitioners, rather than researchers or practitioners. It also requires more continued efforts of small researchers-practitioners to do high-quality research, contribute to the research community, and call attention to their problems.

References

1. Bourne, S. and Cantrill, B. Communications and the practitioner. Commun. ACM 52, 8 (Aug. 2009), 5.
2. Briand, L. Embracing the engineering side of software engineering. IEEE Software 29, 4 (July–Aug. 2012), 96–96.
3. Glass, R.L. One man's quest for the state of software engineering's practice. Commun. ACM 50, 5 (May 2007), 21–23.
4. Hardman L., Obrenovic, Z., and Nack, F., guest eds. Special issue of ACM Multimedia Systems Journal on canonical processes of media production 14, 6 (Dec. 2008), 327–433.
5. Norman, D.A. The research-practice gap: The need for translational developers. interactions 17, 4 (July 2010), 9–12.
6. Obrenović, Ž. and den Haak, B. Integrating end-user customization and authentication: The identity crisis. IEEE Security and Privacy 10, 5 (Sept./Oct. 2012), 82–85.
7. Spector A., Norvig, P., and Petrov, S. Google's hybrid approach to research. Commun. ACM 55, 7 (July 2012), 34–37.
8. Stolterman E. The nature of design practice and implications for interaction design research. International Journal of Design (IJDesign) 2, 1 (2008).

2. The Hawthorne Studies

The Hawthorne Studies are probably one of the most important and often forgotten studies of human factors. This study is still relevant as it illustrates the difficulty of studying realistic complex human issues in realistic situations. It also illustrates that lasting and robust research contributions related to real-world human issues may be based on inquiry from within industry rather than initiated by academia and commissioned by funding bodies.[1]

Aerial view of the Hawthorne Works, ca. 1925.

---

[1]This chapter is based on the article The Hawthorne studies and their relevance to HCI research, interactions 21, 6 (October 2014), 46-51.

Introduction

The Hawthorne studies are best known for the Hawthorne effect, namely that those who perceive themselves as members of the experimental or otherwise favored group tend to outperform their controls, often regardless of the intervention. Secondary sources describing the Hawthorne effect (e.g., [1,2]) tell us that in an experiment conducted at Western Electric's Hawthorne Works factory in the 1920s, psychologists examined the working conditions of plant workers doing repetitive tasks. The major finding quoted is that irrespective of what one does to improve or degrade conditions, productivity goes up. The usual example given is variation in light. If light conditions improved, so did productivity; however, when light conditions were downgraded, productivity again went up.

Unfortunately, this oversimplified story about the Hawthorne effect overshadows the groundbreaking contribution of the Hawthorne studies. The Hawthorne effect is not only controversial [1]—it's also probably the least interesting and least relevant result of this landmark study. The famous light experiment at the Hawthorne plant was just one of more than 30 experiments involving repetitive workers (e.g., relay assemblers, mica splitters) as well as supervisors and other decision makers [3]. And the results were not quite as simple as secondary sources may suggest.

Here, I want to call attention to some of the breakthroughs credited to the Hawthorne studies, which have made a number of practical, conceptual, and methodological innovations in human factors, management studies, and sociology. I also want to argue that even though these studies were performed more than 80 years ago, the HCI research community can still learn something from them.

The Hawthorne Studies

The Hawthorne studies were conducted at Western Electric's large plant outside Chicago. The research outcomes are reported in Roethlisberger and Dickson [4]. I would also recommend Jeffrey Sonnenfeld's detailed analysis of the studies and their influence [5], as well as the online resources at the Harvard Business School [6].

In a period between 1924 and 1933, six studies were performed. These studies were longitudinal in nature, running between several months and several years.

The Illumination Studies (1924). The studies began in 1924 when researchers, together with the National Research Council of the National Academy of Sciences, tried to examine the relationship between light intensity and employee productivity at the Hawthorne Works plant. The expectation was that an increase in lighting would lead to an increase in productivity, and vice versa. But an impressive team of industrial specialists and academics was not able to find any consistent correlation between lighting levels and worker output. The productivity increased with brighter intensity, but also with lower intensity, as well as when researchers only pretended to increase or decrease the intensity of light. No further tests were planned originally, but researchers were puzzled by these unanticipated results. They realized there was not a simple answer to the issue of illumination and worker productivity. The psychological and sociological issues, which were not controlled, presented a major problem with the test results.

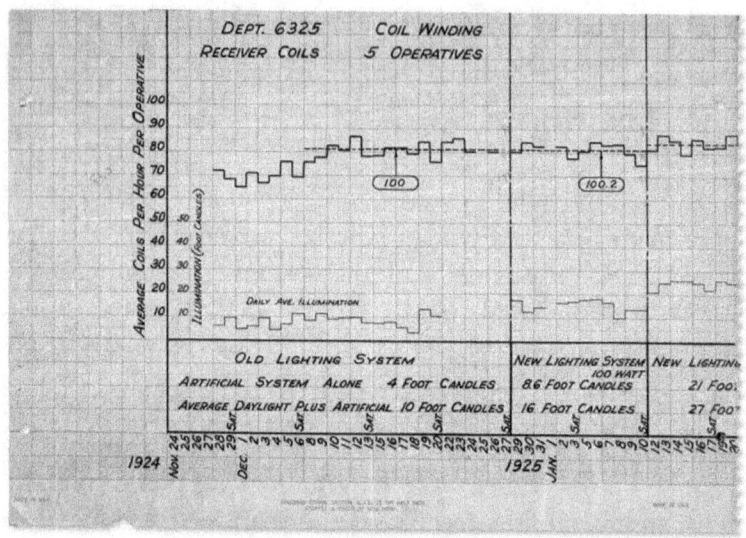

A log from the illumination studies.

At this point, the National Research Council withdrew from the project. However, Western Electric decided to continue studies in collaboration with Harvard University researchers, including Fritz Rothlisberger, W. Lloyd Warner, and Elton Mayo. They also changed the research objective from a study of illumination to a study of the physical factors that cause fatigue and monotony.

The Relay Assembly Test Room (1927–1932). In the next research phase, five workers assembled magnetic relays, working in isolation from the main shop. The separation was chosen to better control working conditions and to more deeply investigate productivity. Researchers collected and analyzed a significant amount of data, including mechanical records of worker output, a daily record of comments made by researchers and study members, observers' logs of work activity, results of periodic medical examinations of workers, and interview transcripts.

This research phase lasted five years, but experimental conditions were maintained for only the first two and a half years. Thirteen experimental treatments took place, including variations in the number and duration of rest breaks, the length of workdays, and the length of workweeks. Throughout the studies, production steadily rose. At the end, when the original, more demanding conditions were reinstated, the productivity of workers dropped only slightly, to a 30 percent increase over the original output. In addition, absenteeism dropped to a third of the original absenteeism. Evidence to support the initial hypotheses about relief from fatigue and monotony was inconclusive.

Women in the assembly test room.

Two derivative studies (1928–1929). Because researchers were unsure what caused the significant increase in productivity in the Relay Assembly Test Room, they launched two derivative studies. These explored two possible explanations for improvements in productivity. One explanation was the different incentive systems. Researchers realized that in the Relay Assembly Test Room, workers in a smaller group could more directly affect their group-based compensation, compared with 200 assembly workers in the main shop. A new relay group with a small-group-output incentive plan was arranged on the shop floor, but without being isolated from the other workers. Productivity quickly increased by 12 percent but leveled off for the duration of the study.

The second derivative study was performed to test the effect of isolation of a small group on productivity. A group of workers did not receive a different incentive plan but were placed in a separate room. Group output increased in the early phase of this project by 15 percent. However, the investigators also realized that external factors had a much more significant effect on productivity than any of their interventions. For instance, when rumors about the possible transfer of some jobs away from Hawthorne appeared, productivity began to drop significantly.

The conclusion from the derivative studies was that the wage incentive had some role in the productivity increase, but that it certainly did not completely explain the productivity increase in the Relay Assembly Test Room. Furthermore, the investigators concluded that it was not possible to identify the independent influence of wage incentives on productivity because it is so intertwined with other variables.

Interview program (1928–1930). The management and investigators were impressed with the great potential of workers if they were given proper conditions. But they were uncertain about what these conditions might be. For that reason, from 1928 to 1930 they interviewed around 21,000 employees. The interviews were then analyzed and classified by the articulated complaints.

Interviewers made a number of interesting findings. But they also quickly discovered that it was not enough to simply catalog complaints. Management was aware of many of the complaints, but out of context, complaints were misleading. However, understanding the personal and economic background of the workers made possible a much richer appreciation of the importance of a given complaint.

The interview program also suggested there was a great motivational value in directly asking workers for their opinions and perceptions and listening closely to their responses, as well as recognizing the relationships between workers' work and non-work lives.

Bank Wiring Observation Room (1931–1932). Hawthorne investigators also observed the relative social positions of different jobholders in a group. In the Bank Wiring Observation Room, the last research phase in the Hawthorne studies, investigators looked at 14 workers in three different jobs. They worked together to produce wired equipment for use in switches. The goal was to investigate the status distinctions and social relations in the workplace. The researchers discovered an unexpected culture, revealed through group norms and activities such as informal leadership patterns, restriction of output, group discipline, friendship, job trading, and cooperation. In this final research phase, the investigators developed hypotheses about the conditions that encourage the creation of an informal culture, which may be either compatible with or hostile to managerial intentions.

The Legacy of the Hawthorne Studies

It is easy to see the Hawthorne studies as a failure. None of the findings obtained were very conclusive, and the studies were also imperfect from a methodology point of view. Critiques were often harsh, with results being called "injudicious," "scientifically worthless," "the myths of Hawthorne," and the result of "cow sociologists" [5].

Sonnenfeld, however, noted that many critiques of the Hawthorne studies were incorrect and out of context, claiming that *"the gunsmoke of academic snipers can obscure the conceptual contribution of these pioneering efforts"* [5]. He elaborated that the Hawthorne studies were conducted in a manner that led not to the testing of theories, but to their development. Consequently, their greatest contribution was to expand the concepts of organizational behavior beyond Frederick Taylor's notion of scientific management. At that time the prevailing view was that people went to work purely for money and to earn a living. The Hawthorne studies showed convincingly that this view was deeply flawed. *"Instead of treating the workers as an appendage to 'the machine,'"* Sonnenfeld noted, the Hawthorne studies brought to light ideas concerning motivational influences, job satisfaction, resistance to change, group norms, worker participation, and effective leadership [5]. In the 1930s, these were groundbreaking concepts. Under the influence of the Hawthorne studies, management teaching and practice changed significantly. The Hawthorne research stimulated thought on individual differences and job matching, work design, incentive plans, employee participation, the social nature of organizational activities, small work groups, and leadership. The findings from the studies have been credited with contributing to the later development of social science topics, including small-group behavior, client-centered theory, organization theory, and research methodology.

Porter, Lawler, and Hackman noted that the Hawthorne investigators were the first to highlight the social complexities of organization life:

"From the time of the publication of the results of the Hawthorne Studies onward, no one interested in the behaviour of employees could consider them as isolated individuals. Rather, such factors and concepts as group influences, social status, informal communication, roles, norms, and the like were drawn upon to explain and interpret the voluminous data from these studies and other field investigations that followed them." [7].

As a consequence, in social and management research, the study of static social structures practically disappeared after the publication of the Hawthorne research.

The Hawthorne studies significantly contributed to the development of research methodology for studying complex social situations. Hawthorne investigators were initially convinced that controlled experiment was the best methodology for their research. Through the studies they made a significant methodological shift, in which they recognized the impossibility of applying the controlled experiment approach for the questions they were addressing. Roethlisberger and Dickson summarize this shift:

The difficulty, however, went much deeper than the personal feelings of failure of the investigators. They were entertaining two incompatible points of view. On the one hand, they were trying to maintain a controlled experiment in which they could test for the effects of single variables while holding all other factors constant. On the other hand, they were trying to create a human situation, which remained unaffected by their own activities. It became evident that in human situations not only was it practically impossible to keep all other factors constant, but trying to do so in itself introduced the biggest change of all; in other words, the investigators had not been studying an ordinary shop situation but a socially contrived situation of their own making.

With this realization, the inquiry changed its character. No longer were the investigators interested in testing for the effects of single variables. In the place of a controlled experiment, they substituted the notion of a social situation, which needed to be described and understood as a system of interdependent elements* [4].

Individual human behavior is determined by a complex set of factors and is rarely a consequence of a simple cause-and-effect relationship.

The Relevance of the Hawthorne Studies for Computer Science Research

Those who cannot remember the past are condemned to repeat it.—George Santayana

The story of the Hawthorne studies is in many ways similar to the story about the development of the many computer science (CS) fields. Liam Bannon, for example, observed that the HCI (human-computer interaction) discipline has moved from early studies of human factors and experimental evaluation of interfaces toward the general sense-making of our world:

The area of concern [of HCI] is much broader than the simple "fit" between people and technology to improve productivity (as in the classic human factors mold); it encompasses a much more challenging territory that includes the goals and activities of people, their values, and the tools and environments that help shape their everyday lives [8].

In other words, HCI is much more than efficient user data input or output. Solutions to HCI problems do not reside in simple ergonomic corrections to user interfaces. Hawthorne investigators drew similar conclusions about working conditions, namely that simple ergonomic corrections, such as changing the light intensity, are not sufficient to improve productivity and are certainly not the most significant factors that influence productivity.

Another reason the Hawthorne studies are relevant for CS research is methodological. Investigators made a shift from controlled experiments toward approaching a complex social situation as a system of interdependent elements. In many ways, this shift is what we experience today in HCI and interaction design communities. Bannon, for example, argued that the introduction of the computer supported cooperative work (CSCW) field presented "*a shift from a psychological to a sociological perspective on human work and activity, emphasizing field observation methods rather than lab studies*" [8]. The Hawthorne studies provide an illustration of why such approaches are needed when studying complex human and social phenomena.

Furthermore, the studies showed the value of careful observation and honest reporting of research failures and successes. As noted by Jonathan Arnowitz and Elizabeth Dykstra-Erickson, "[t]he great value of the Hawthorne experience is in learning to observe and keep on observing, especially when an initial *causal relationship doesn't quite account for the observed interaction*" [2]. In many situations Hawthorne researchers were confused, and they admitted it. But they continued to carefully observe and document all of their findings. The original elaborate report of the studies, Management and the Worker [4], is a model of honest reporting of research. It describes, in a chronological order, the things investigators did, the judgments they made, the leads they followed, and the conclusions they drew. Roethlisberger and Dickson selected this method to "*picture the trails and tribulations of a research investigator at his work and thus allow future investigators to see and profit from the mistakes which were made*" [4]. This approach makes Management and the Worker, even after 75 years, a relevant and surprisingly insightful book, useful for anyone who wants to understand the difficulty of studying realistic complex human issues in realistic situations. I would recommend it as standard reading for CS researchers.

The Hawthorne studies also demonstrated the value of doing research in practice, over a long period, and with real users and realistic tasks. A related issue is the fact that the Hawthorne studies produced useful results primarily because of the interest and support of Hawthorne Works. While researchers from the academy were involved, the main initiative did not come from the academic side or from funding agencies (the National Research Council of the National Academy of Sciences withdrew after the initial "failure" of the illumination test). I think this may be an important lesson for the HCI community. It suggests that lasting and robust research contributions related to real-world human issues may be those based on inquiry from within industry rather than those initiated by academia and commissioned by funding bodies.

Lastly, the Hawthorne studies illustrated that the value of

research is not necessarily derivation of conclusive results. The legacy of these studies is a realization that treating the workers as an "appendage to 'the machine'" with the goal of improving the human-machine "fit" is a flawed conceptual framework [5]. This legacy may stimulate us to look differently at some HCI contributions. Similar to the Hawthorne studies, the lasting impact of some HCI research may be not the results of laboratory experiments, but rather an expansion of the concepts of HCI beyond notions of human-computer "fit" and the identification of new concepts that can help us to understand human activities mediated by computing.

References

1. Macefield, R. Usability studies and the Hawthorne Effect. Journal of Usability Studies 2, 3 (2007), 145–154.
2. Arnowitz, J. and Dykstra-Erickson, E. Observation and interaction design: Lessons from the past. Interactions 14, 6 (Nov. 2007), 64–ff.
3. Brown, A.L. Design experiments: Theoretical and methodological challenges in creating complex interventions in classroom settings. The Journal of the Learning Sciences 2, 2 (1992), 141–178.
4. Roethlisberger, F.J. and Dickson, W.J. Management and the Worker. Harvard University Press, 1939.
5. Sonnenfeld, J.A. Shedding light on the Hawthorne Studies. J. Occupational Behavior 6, 2 (1985), 111–130.
6. http://www.library.hbs.edu/hc/hawthorne/
7. Porter, L.W., Lawler, E.E., and Hackman, J.R. Behaviour in Organizations. McGraw-Hill, New York, 1975.
8. Bannon, L. Reimagining HCI: Toward a more human-centered perspective. Interactions 18, 4 (2011), 50–57.

The last remaining portion of Western Electric's Hawthorne Works factory.

3. Is Academia Guilty of Intellectual Colonization of Practice?

Creating partnerships between academia and industry is valuable but challenging. Doing research in practice is rare and difficult. But doing it naively creates more harm than good. The following essay is my reaction to negative side-effects of naive academia-industry partherships.[1]

Plato's academy mosaic. Wikimedia Commons.

---

[1] This chapter is based on my BLOGS@CACM post Is Academia Guilty of Intellectual Colonization of Practice?

30 Is Academia Guilty of Intellectual Colonization of Practice?

Due to my position between industry and academia, over the past few years I've been reading a lot and writing a bit about the research-practice gap in computer science (CS). I've discovered that many of the issues about the research-practice gap in CS are not unique to our field. Other fields, such as medical or social sciences, have been facing similar problems. I believe we can learn a lot from them.

Recently, I came across an interesting article about the research-practice gap in health services. In the editorial *"Research in general practice: who is calling the tune*[2]*,"* Tom O'Dowd made the following claim:

"Often in health services research the questions are asked and answered by people outside general practice but using general practitioners as respondents or data gatherers. There is usually no involvement of the general practitioners in the analysis or discussion of the project, although their time is acknowledged. Surely this is a modern form of colonization at intellectual and professional levels." (British Journal of General Practice, October 1995, page 515)

After reading this editorial, I started to wonder if this claim would be true in my own field. So I did a simple exercise, and replaced "general practice" with "software engineering" in Tom O'Dowd's original text:

"Often in software engineering [or another CS field] research the questions are asked and answered by people outside software engineering practice but using software engineering practitioners as respondents or data gatherers. There is usually no involvement of the software engineering practitioners in the analysis or discussion of the project, although their time is acknowledged. Surely this is a modern form of colonization

---

[2] https://www.ncbi.nlm.nih.gov/pmc/articles/PMC1239398/

at intellectual and professional levels."

I am of the opinion that the adapted claim is true, at least partially. I have seen many research contributions about practice where I felt that involved practitioners could have been credited more or could have given an opportunity to contribute as co-authors. As a practitioner, I also have been approached a number of times to participate in research projects where my role was limited to anonymously filling in questionaries' and interview forms. None of the times I was invited or given an opportunity to actively participate in analysis or discussions. The researchers had good intentions. But they were amateurs, usually students with little or no practical experience in domains they were studying.

My main reaction in mentioned situations was not the feeling of being "colonized." Rather, it is the feeling of wasted effort and missed opportunity. Doing research in practice is rare and difficult, but doing it naively creates more harm than good. At best, it does not help bridging the research-practice gap. At worst, it makes the gap wider, as practitioners may be reluctant to collaborate in any further activities.

I would definitely like to encourage academic researchers to connect to practice more. But academic researchers should be careful not to be perceived as "intellectual colonists." They should not treat practitioners as mere objects of their studies or simple sources of data. In my view, default mode of collaboration between researchers and practitioners should be a research partnership. Practitioners can provide invaluable real-world knowledge and experience. As noted by Fred Brooks, it is easy for academic researchers to overlook some crucial properties of real-world (The Design of Design, page 82). Practitioners, on the other hand, generally lack research skills. Academic researchers, therefore, can help creating the research climate in practice. They can supervise practitioners in doing research, and further expand or generalize initial research results and connect the results to existing body of knowledge.

I am happy to see that several venues are providing space for research-practice partnerships. The IEEE Software Insights column, for instance, publishes contributions from software engineering practitioners, offering them coaching and mentoring, informal reviews prior to submission, an official peer review, and professional editing support. The CACM Practice column and ACM Queue follow similar paths. But, in general, this type of collaboration is unusual, and not encouraged in academia.

Academic researchers are not the only ones to be blamed. Practitioners should not let themselves be "colonized." Practitioners are often uninterested in research and passive in interaction with researchers. To benefit from research projects and collaboration with academia, practitioners need to be more proactive. They need to propose research questions, do some research themselves, and proactively look for opportunities to create partnerships with academic researchers.

As a take-away message, I propose introducing the "anti-colonization test" for projects that involve collaboration between academic researchers and practitioners. This test requires repeating the same exercise I did in the beginning: taking Tom O'Dowd's claim about general practice research and replacing the term "general practitioner" with "my own field practitioner." If the resulting claim even remotely looks as something that may be true, then the project has failed the test. It may a good moment to take a step back and rethink the proposed researcher-practitioner collaboration form.

4. The Four Points of the Research Compass

Doing research is difficult. Doing research in practice even more. In this essay I explore what motivates and drives researchers to do research.[1]

The Flight Director Attitude Indicator (FDAI) of the Apollo Guidance Computer (AGC). Wikimedia Commons.

---

[1] This chapter is based on the article The four points of the HCI research compass, interactions 20, 3 (May 2013), 34-37.

Introduction

Most discussions about research in computer sciences (CS) focus on research methods and skills, on the question of how research should be conducted (e.g., [1]). Research tools and methods, however, are only passive instruments in the hands of motivated researchers. But what exactly motivates and drives CS researchers? Here I address this question, arguing that we need to be more thoughtful about our research motivation, not just our research skills.

Research skills can help us to do the research properly, but research motivation is the main force behind all of our research efforts. To support this discussion, and inspired by recent ideas from philosophy [2], I use the metaphor of a compass to discuss research motivation on the meta level, independent of the research methods being used. Using this metaphor I present a new, higher-level view on CS research as being driven by four main motivators (Figure 1):

- **Curiosity and wonder**, where we follow our strong interests and desires to learn new things;
- **Doubt**, where we want to obtain deeper and more certain understanding;
- **Belief and vision**, where we set or follow research ideals; and
- **Skepticism**, where we question the possibility of reaching some research goals.

I argue here that these four directions are legitimate motivators for doing research, and that we need to support efforts in all of them. Indeed, each of these motivators has positive and negative sides, and awareness of the pros and cons can help us to do better research. I also contend the main question that we as a community need to answer is not which of these directions to follow, but rather, what is the right balance among contributions motivated by all four sides. I illustrate each of these motivators with concrete examples from the CS field.

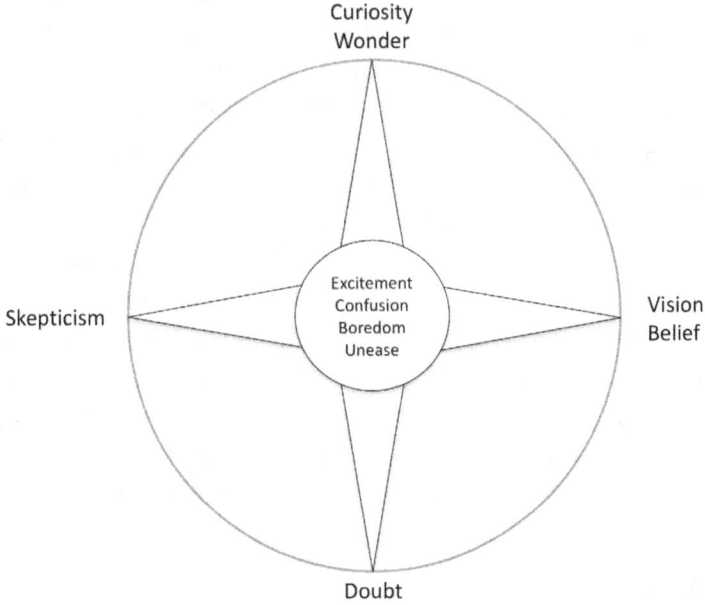

Figure 1. The four points on the research compass.

Curiosity and Wonder

As the simplest and the most obvious point of the research compass, curiosity and wonder describe the natural characteristic of researchers to have a strong interest in and eagerness to know more about a topic. Many CS research contributions came from researchers being fascinated by or curious about some issue, including technology, people, or interaction between technology and people. Curiosity and wonder are also closely connected to academic freedom. Brad Myers emphasized the pivotal role of creative and curious university researchers in the advancement of the human-computer interaction (HCI) field [3].

While curiosity and wonder are the driving force behind innovations, they alone are not sufficient for a research contribution. If overemphasized, they may even produce negative effects. For instance, they may prevent us from obtaining more depth in our research. As noted by Saul Greenberg and Bill Buxton, the CHI conference sometimes favors innovative and more "curious" solutions at the price of more "doubting" ones [4]. They argued that contributions that reexamine existing results are seen as "replications," non-original contributions that are not valued highly, and when they are reviewed the typical referee response is "it has been done before; therefore, there is little value added." In addition, as we are working in a discipline that studies people, we may need to restrain our curiosity due to a number of sensitive ethical issues.

Doubt

When we make some discovery we may ask ourselves if our findings are wrong, coincidental, or a result of wishful thinking. Such questions are the beginning of doubt, one of the most important motivators behind research. Any evaluation can be viewed as an effort to reduce doubts about our findings. Experimental evaluation, for example, doubts claims in the form of a null hypothesis, a claim that an observed phenomena has nothing to do with our intervention. We need to invest effort to disprove the null hypothesis, thereby reducing the doubt that our findings are accidental. Ethnographic methods and techniques, such as protocols for interviews, recordings, and analytic frameworks, such as grounded theory, add rigor and discipline to studying complex social phenomena. In this way, they are forcing us to systematically document and analyze observed phenomena, adding certainty to our observations and conclusions. Similarly, by publishing our articles in peer-reviewed venues, we subject our findings to the doubt of expert peers.

As a research motivator, doubt is primarily a positive force. Contrary to skepticism, doubt does not question the possibility of knowing something or the validity of pursuing some direction. When we doubt some finding, we want to set it on firmer ground and add more certainty to it. Shumin Zhai nicely argued for this in his discussion of the importance of evaluation in HCI:

"It is the lack of strong theories, models and laws that force us do evaluative experiments that check our intuition and imagination. With well-established physical laws and models, modern engineering practice does not need comparative testing for every design. The confidence comes from calculations based on theory and experience. But lacking the ability to do calculations of this sort, we must resort to evaluation and testing, if we do not want to turn HCI into a 'faith-based' enterprise" [5].

Too much doubt also has its disadvantages; for example, it can lead to situations in which we work only on minor improvements that can be easily tested, but do not produce enough innovation. This topic has been a subject of discussion in the CHI community for years (e.g., [4,5]).

Vision and Belief

While the term belief may have a negative connotation in the scientific world because of its vague definition and association with religion, it is difficult to imagine any research activity without some form of belief or guiding vision. We normally believe in the importance of doing research in our domain (and hope that funding agencies share our belief) even without having strong evidence about the value of doing such research (yet). The value of "fundamental research," for example, may become evident only after a long period, if ever. In the field of HCI, Buxton talked about the long cycle of innovation, noting that it may take several decades for a research innovation to become valuable in practice:

"The move from inception to ubiquity can take 30 years.... The first prototype of a computer mouse appeared as a wooden box with two wheels on it in the early 1960s, about 30 years before it achieved the level of 'ubiquity'" [6].

Looking at the larger scale, behind many subfields of computer science and HCI we may find a few visionary contributions that have driven and inspired other researchers. Many core ideas in HCI are inspired by Vannevar Bush's "memex" paper [7], J.C.R. Licklider's vision of networked IT in the 1960s, and Douglas Engelbart's NLS (online system) demonstration at the Fall Joint Computer Conference in San Francisco in December 1968 [8]. Douglas Engelbart received the ACM Turing Award in 1997 for *"an inspiring vision of the future of interactive computing and the invention of key technologies to help realize this vision"*. Don Norman's book The Psychology/Design of Everyday Things practically defined a new domain—user-centered design—inspiring thousands of HCI contributions (Google Scholar citation count close to

10,000) [9]. Similarly, Mark Weiser's article "The Computer for the 21st Century" has been an inspiration for thousands of contributions (Google Scholar citation count above 9,000) [10]. Visionary contributions can also be a result of a community effort. Communications of the ACM, for instance, published a number of special issues introducing a shared vision for many of the HCI subfields, including perceptual UIs (2000), attentive UIs (2003), and organic UIs (2008). Similar roles may be played by workshops or events such as Dagstuhl Seminars. Such initiatives serve an important role in outlining or consolidating a new field, defining its basic terminology, and setting a high-level research agenda.

Vision is a very important component of any community effort, as shared vision can inspire researchers and enable synergic development of the field. Such vision can also come from the outside—for example, from funding agencies. The European Commission (EC), within its Framework Programs, defines themes and "challenges," such as "pervasive and trusted network and service infrastructures" and "learning and access to cultural resources," which are used to guide and prioritize the funding of research projects. Similarly, the U.S. National Science Foundation (NSF) has a number of core programs aimed at stimulating and guiding research in particular directions, such as "human-centered computing" and "robust intelligence."

On an individual level, having a personal vision can help us to define our research line and identity. A common component of the academic job application, for example, is the research statement, in which applicants are expected to express the future direction and potential of their work and propose a valuable, ambitious, but realistic research agenda. For personal development, it is important to continuously work on the personal vision, and to make it explicit and open to critical reflection and discussion with mentors and colleagues.

Too much reliance on the vision, on the other hand, may have some negative consequences. While vision may inspire and guide research, vision makes sense only if it is followed by a number of curious and doubting contributions. If we get too excited about the vision we are following, we may become less critical about our findings. This can lead to confirmation bias, a tendency to favor information that confirms our beliefs or hypotheses. Vision can guide us in the wrong direction. We may also end up with visions that are too narrow. This may lead to overspecialization and to situations in which we are blind to innovative solutions because they are beyond the scope of any of the currently active visions. In addition, to be useful, vision should be based on deep knowledge and understanding of the research field, not on its ignorance.

Vision and belief are much more complex research motivators than curiosity and wonder. When we are driven by curiosity, we simply follow interests and the desire to learn something new. Vision and belief, on the other hand, require longer-term commitment to some idea, as well as constant effort to focus and organize research activities.

Skepticism

Skepticism is a loaded term with a number of definitions. Closest to the meaning I use here is the definition of skepticism as "doubt regarding claims that are taken for granted elsewhere" [11]. I view research skepticism in a similar fashion, as a reality checker that questions the fundamental premises we normally take as a given. As such, skepticism can call attention to the viability, feasibility, or practicality of a research direction or approach. Contrary to doubt, which can motivate us to further investigate some topic to obtain more certainty, skepticism may call us to abandon some line of inquiry and consider alternatives.

Fred Brooks's paper "No Silver Bullet—Essence and Accidents of Software Engineering" is probably one of the best examples of useful skeptical thought in computer science [12]. Brooks expressed his skepticism toward approaches to software engineering research that aim to discover a single solution that can improve software productivity by an order of magnitude. Brooks seriously questioned the possibility of ever finding such "startling breakthroughs," arguing that such solutions may be inconsistent with the nature of software. Brooks also made clear that his skepticism is not pessimism. While Brooks questioned the possibility of finding a single startling breakthrough that will improve software productivity by an order of magnitude, he believed that such improvement can be achieved through disciplined, consistent effort to develop, propagate, and exploit a number of smaller, more modest innovations. In "Human-Centered Design Considered Harmful," Norman was skeptical about naive approaches to human-centered design (HCD), stating that HCD has become such a dominant theme in design that interface and application designers now accept it automatically, without thought, let alone criticism [13]. The Greenberg and Buxton paper "Usability Evaluation Considered Harmful (Some of the Time)" provides a similar skeptical view on the HCI practice, encouraged by educational institutes, academic review processes, and institutions with usability groups, which promote usability evaluation as a critical part of every design process [4]. Based on their rich experiences, they argued that if done naively, by rule rather than by thought, usability evaluation can be ineffective and even harmful.

Skepticism can be a useful antidote to too much excitement or opportunism in doing research. Skeptical contributions, if well argued, can prevent the wasting of energy and resources in pursuing wrong directions and stimulate us to rethink our approach. The same applies on the individual level. Having curious and enthusiastic students guided by experienced and more skeptical mentors is a proven and very successful model for educating researchers.

Too much skepticism, on the other hand, comes with negative side effects. Chris Welty nicely described this problem as what he called "unimpressed scientist syndrome." In his keynote speech at the 2007 International Semantic Web Conference, Welty portrayed his personal history of strong skepticism toward many computing innovations that later become very successful, including email, the World Wide Web, and the Semantic Web [14]. He argued that this may be a wider problem, and that many academic researchers are skeptical by rule rather than by thought, rejecting innovative solutions without serious consideration with phrases such as "I've seen this before; this is not gonna work." Furthermore, if skepticism is not a well-argued result of long experience, it may trigger an emotional debate without contributing much to it. Similar to vision, useful skepticism requires deep knowledge and a fundamental understanding of the research field. It is not surprising that many skeptical authors are also authors of influential visionary contributions.

Skepticism is probably one of the most complex influences on research. Contrary to doubt, which can rely on a number of tools and methods (e.g., experiments, ethnographical methods), there are no simple and structured tools for skepticism. Useful skepticism requires careful thought, experience, and an excellent overview of the field.

Conclusion

The four points of our research compass metaphor do not suggest that research contributions should be motivated by only one direction. Even individual contributions usually combine several elements, presenting our discoveries (curiosity), for instance, with their evaluations (doubt). In a team it is good to have individuals with different affinities. At a community level, it is equally important to have contributions motivated by all four points. The community cannot develop without new ideas and new visions, but without a healthy dose of doubt and skepticism, we can get incorrect results or go in a faulty direction. It is also the responsibility of the community to set high standards and maintain the right balance among contributions originating from different research motivations.

I hope the research-compass metaphor can help researchers to be more thoughtful about their professional development and stimulate them to ask themselves questions such as:

Are we curious enough about topics of our research? Do we explore enough or do we jump too quickly to tests? Do we have a plan to maintain our curiosity, such as a sabbatical leave? * Do we have vision about where we would like to go, or are we simply following the latest trends? * Do we doubt our findings enough and are we using the right methods? * Are we skeptical enough about our own work? Are we too skeptical as reviewers? Are we more skeptical toward some contributions and less toward others?

I believe that answering these questions can make us more thoughtful about our motivation and enable us to make more informed decisions about our development as researchers, as well as about the development of our research field.

References

1. Lazar, J., Feng, J.H., and Hochheiser, H. Research Methods in Human-Computer Interaction. John Wiley & Sons, 2010.
2. Alexander, J. The four points of the compass. Philosophy 87, 1 (January 2012), 79–107.
3. Myers, B.A. A brief history of human-computer interaction technology. interactions 5, 2 (March 1998), 44–54.
4. Greenberg, S. and Buxton, B. Usability evaluation considered harmful (some of the time). Proc. of CHI '08. ACM, New York, 2008, 111–120.
5. Zhai, S. Evaluation is the worst form of HCI research except all those other forms that have been tried. 2003;[2]
6. www.businessinnovationfactory.com/iss/innovators/ bill-buxton[3]
7. Bush, V. As we may think. The Atlantic Monthly 176, 1 (July 1945), 101–108.
8. Canny, J. The future of human-computer interaction. Queue 4, 6 (July 2006), 24–32.
9. Norman, D.A. The Design of Everyday Things. L. Erlbaum Assoc., Inc., Hillsdale, NJ, 1988.
10. Weiser, M. The computer for the 21st century. Scientific American 265, 3 (September 1991), 94–104.
11. http://en.wikipedia.org/wiki/Skepticism
12. Brooks, F.P. No silver bullet—Essence and accidents of software engineering. IEEE Computer 20, 4 (April 1987), 10–19.
13. Norman D.A. Human-centered design considered harmful. interactions 12, 4 (July 2005), 14–19.
14. http://videolectures.net/iswc07_welty_hiwr/

---

[2]http://www.shuminzhai.com/papers/EvaluationDemocracy.htm
[3]http://www.businessinnovationfactory.com/iss/innovators/bill-buxton

5. Insights from the Past

Research community mostly aims at creating ground-breaking and innovative solutions. Looking back at older and historical developments is rarely considered interesting or relevant. However, in the domain of software engineering, lessons from the past are still actual. That was one of the reasons for me to start to work on creating the website that makes it easier to explore the rich history of software engineering. In this chapter I reflect on lessons learned, and the value of studying the past so that it might not be reinvented but become a source of inspiration for the present and future.[1]

A snapshot from IEEE Software history site, showing covers from 1980s.

---

[1]This chapter is based on the article Insights from the Past: The IEEE Software History Experiment, IEEE Software 34, 4 (July-August, 2017), 71-78.

Introduction

The IEEE Software history website[2] is a curated site complementing the official IEEE Software website[3]. It offers a look at IEEE Software's history at a glance. Here, I discuss its genesis' how it illustrates the practical value of historical data, and how it offers a glimpse into the magazine's future.

The Website's Genesis

The website has developed organically. There never was an official project to develop an elaborate overview of the magazine's history. Rather, the website evolved somewhat out of curiosity and as a volunteer initiative, partly in reaction to positive feedback about its developing content.

The idea for the website arose during the 2016 IEEE Software editorial-board meeting at the Software Improvement Group (SIG) in Amsterdam. As an organizer of the meeting, I was looking for ways to create an IEEE Software atmosphere. I decided to print the magazine covers and put them up as wallpaper (see Figure 1).

The board members, who were from both academia and industry, liked such an overview of topics and trends that were once considered important. For many, it brought back memories or created awareness of missed topics. In addition, the covers are attractive. SIG kept them on the wall for several months after the meeting.

After receiving requests to share digital versions of the covers, I created a simple website to display them. Thus, the original idea of the history website was only to create that display.

---

[2] obren.info/ieeesw
[3] www.computer.org/software

Figure 1. IEEE Software front covers displayed at the 2016 editorial-board meeting.

The Website's Content

While collecting the covers, I discovered that the July/August 2017 issue would be the 200th issue. So, I decided to use the history website to celebrate this anniversary. I extended the website with several types of content, including these:

- *More than 1,000 quotes.* This was the most rewarding part of creating the site. These curated quotes make the website much more than a simple metadata index. The quotes have also been important in creating interesting content to promote the history of IEEE Software on social media such as Twitter because they're short but informative.
- *Indexes of all 3,000+ articles and 4,000+ authors.* These indexes enable quick exploration of articles and authors in a historical context. I also added a historical timeline for search results.
- *A citation index* (based on Google Scholar searches) correlated with the publication year. This helps show IEEE Software articles' broader impact. It also aids identifying the most cited articles, authors, or themes.

For more on the website's content, see the sidebars.

Historical Data's Practical Value

Although, at the time of writing of this artcile, the history website was just six months old, I had gathered enough experience to reflect on its value. I classify these lessons learned into the following categories.

Seeing Trends

Historical data enables us to see trends in software engineering research and practice. In many aspects, this is the history website's main value, compared to digital libraries such as the IEEE Computer Society Digital Library (CSDL; www.computer.org/csdl) and IEEE Xplore (ieeexplore.ieee.org).

To illustrate the possibilities of seeing trends, Figure 2 shows word clouds created from terms in IEEE Software article titles, for four decades. Although IEEE Software covers diverse topics, each decade has had a few topics that were more popular.

In the 1980s, topics related to different programming paradigms were popular. For instance, parallel and distributed programming were important topics. Six theme issues and cover articles discussed programming:

- 1984, no. 2. Programming: Sorcery or Science?
- 1986, no. 4. Firmware Engineering: The Interaction of Microprogramming and Software Technology.
- 1988, no. 1. Parallel Programming: Issues and Questions.
- 1988, no. 3. What Is Object-Oriented Programming?
- 1989, no. 4. Parallel Programming: Harnessing the Hardware.

- 1989, no. 5. A Compositional Approach to Multiparadigm Programming.

In the 1990s, the focus shifted toward process-related topics. Measurements, metrics, and quality assurance also received significant attention. Ten theme issues covered process management and metrics (and their combination):

- 1990, no. 2. Using Metrics to Quantify Development.
- 1991, no. 4. Process Assessment.
- 1992, no. 4. Reliability Measurement.
- 1993, no. 4. The Move to Mature Process.
- 1994, no. 4. Measurement-Based Process Improvement.
- 1996, no. 4. Managing Large Software Projects.
- 1997, no. 2. Assessing Measurement.
- 1997, no. 3. Managing Risk.
- 1998, no. 4. Menace or Masterpiece? Managing Legacy Systems.
- 1999, no. 2. Metrics for Small Projects.

The 2000s were clearly the age of requirements engineering. Overall, IEEE Software has published 163 articles with "Requirements" in the title. Of those articles, 91 (56 percent) were published in the 2000s. Seven theme issues covered requirements engineering:

- 2000, no. 3. Requirements Engineering: Getting the Details Right.
- 2003, no. 1. RE 02: A Major Step toward a Mature Requirements Engineering Community.
- 2004, no. 2. Practical Requirements Engineering Solutions.

- 2005, no. 1. Innovation in Requirements Engineering.
- 2006, no. 3. RE 05: Engineering Successful Products.
- 2007, no. 2. Stakeholders in Requirements Engineering.
- 2008, no. 2. Quality Requirements.

In the 2010s, the focus shifted toward architecture:

- 2010, no. 2. Agility and Architecture.
- 2013, no. 2. Twin Peaks of Requirements and Architecture.
- 2013, no. 6. Architecture Sustainability.
- 2015, no. 5. Software Architecture.
- 2016, no. 6. The Role of the Software Architect.

Overall, IEEE Software has published 167 articles with "Architecture" or "Architect" in the title. Of those articles, 97 (58 percent) have been published since 2010.

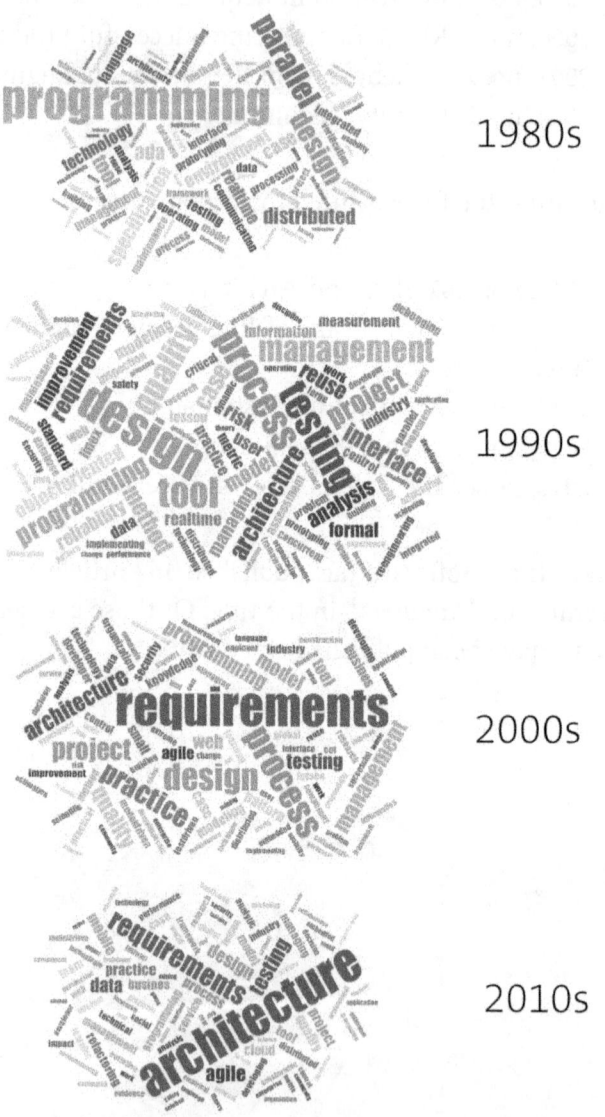

Figure 2. Word clouds with terms in the titles of IEEE Software articles. 1980s. 1990s. 2000s. 2010s.

Preventing Knowledge Inflation

There's a lot of forgetting, and a lot of "never knew that" in our field today [1]. —Robert Glass Those who cannot remember the past are condemned to repeat it. —George Santayana

Knowing history can help us avoid repeating errors and building on each other's work. As Robert Glass noted, we keep forgetting early contributions and often reinvent the wheel. In my experience, much current software engineering work, especially practitioner's books and posts, aren't well connected to previous work. Similarly, Martin Fowler talked about semantic diffusion, which occurs when a definition gets spread through the wider community in a way that weakens it [2]. Often this weakening is a consequence of lack of awareness of the original work related to the definition.

I call this problem the inflation of software engineering terms and knowledge (see Figure 3). The difficulty of finding previous work often leads to reinvention of concepts and solutions. The reinvented solutions get documented and published, leading to the invention of new terms and creation of isolated content (unconnected to previous work). These new terms and content increase the already significant number of articles and posts, which makes finding previous work even more difficult. New cycles of such reinventions are inevitable.

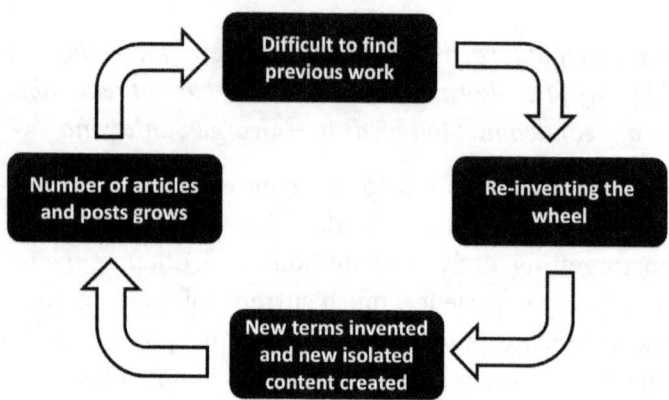

Figure 3. The vicious cycle of inflation of software engineering terms and knowledge. New cycles of such reinventions are inevitable.

Unfortunately, IEEE Software is partly to blame for this cycle. Just by looking at the covers, you can see that many themes repeat. For instance, there have been four "business of software" issues:

- 2002, no. 6. The Business of Software Engineering.
- 2004, no. 5. The Business of Software Engineering.
- 2011, no. 4. Software as a Business.
- 2016, no. 5. The Business of Software.

Looking at the introductions of the later theme issues, you can see that none of them connects to any of the previous ones. Nor do they relate to the brilliant but largely forgotten 1984 issue (no. 3) on Capital-Intensive Software Technology.

IEEE Software provides content that can help root new contributions in previous solid peer-reviewed research and practices. Many IEEE Software authors have been the originators of nowadays mainstream concepts and ideas. For examples, 10 of the 17 authors of the "Manifesto for Agile Software Development"3 have written for IEEE Software: Kent Beck, Alistair Cockburn, Ward Cunningham, Martin Fowler, James Grenning, Andrew Hunt, Ron Jeffries, Robert Martin, Steve Mellor, and Dave Thomas.

Detailed, easily accessible historical data can help slow down the inflation of knowledge by making it easier to find and connect to previous work and ideas.

Being a Source of Inspiration

Another value of easily accessible historical information is in its relevant and inspirational content. I was surprised to discover that many articles from the 1980s and 1990s are still relevant. For example, consider this quote from Bruce Shriver's introduction of the first IEEE Software issue in 1984:

Many of the challenges facing the software industry today are a direct result of our insatiable appetite for new computer-based systems applications. Others confront us simply because we have not managed to successfully solve a large number of problems that we ourselves created many years ago. Specifically, we still, by and large, lack the necessary methods to increase our ability to design and implement high-quality systems [4].

This quote still accurately summarizes current software engineering challenges.

I've come across quite a few such insightful pieces. For my daily work, I've found early work on software architecture, quality, and maintenance still insightful and inspiring. Here are three of my favorite quotes:

Architecture is not so much about the software, but about the people who write the software. The core principles of architecture, such as coupling and cohesion, aren't about the code. The code doesn't "care" about how cohesive or decoupled it is; if anything, tightly coupled software lacks some of the performance snags found in more modular systems. But people do care about their coupling to other team members [5]. We are so used to the notion that quality must take a back seat to productivity that we continue to put up with practices that we know will produce software of lesser quality [6]. The greater speed of technical change means that capital investment must be recovered more quickly and that enhancement and evolution consume proportionately more resources than in a slowly changing technology. This contributes to the fact that maintenance and enhancement are the dominant costs in the software life cycle today [7].

I particularly like the clarity of definitions and research questions in the early articles, which often have defined a new field. Knowing such early work also helps you have authority in the field.

In my practical work as a consultant, I've discovered the value of historical content as an antidote to hype. Nothing cools down a heated sales pitch about a "revolutionary" new technology more than showing the presenter a 30-year-old article describing the same or a similar concept, sometimes with empirical studies, and asking how the "new" solution differs. I've used this tactic successfully a few times. For example, people presenting a new low-code platform are often proud of the platform's use of visual programming that supposedly implements a new programming paradigm. However, as Shi-Kuo Chang's 1987 survey on visual languages shows, many such visual-programming techniques are more than 30 years old.8

I also came across many inspirational but less known and unexpected pieces, such as great articles from Alan Kay and Christopher Alexander:

You could ... say that the main business of everyone on earth is to help everyone else—including ourselves—get enlightened because the technology is getting more and more dangerous [9]. What I am proposing ... is a view of programming as the natural, genetic infrastructure of a living world which you/we are capable of creating, managing, making available, and which could then have the result that a living structure in our towns, houses, work places, cities, becomes an attainable thing. That would be remarkable. It would turn the world around, and make living structure the norm once again, throughout society, and make the world worth living in again [10].

And this just scratches the surface. Please explore these quotes yourself, and use social media to let everyone know when you find some new inspirational pieces.

Another piece of inspiration is what I call "the art of IEEE Software." The covers, as well as the article illustrations, depict key software engineering concepts in an original and artistically pleasing way.

Having Intrinsic Historical Value

History has value in itself. People care about it. For example, Alison Gopnik explained that acknowledging the truth about the past, good or bad, individually or collectively, is deeply important to us as humans, even when it has no immediate effect on the present [11]. I think the same concept applies to the history of software engineering. Many of us software engineering professionals find it important to acknowledge the truth about the past for its own sake, even when it has no immediate effect on what we do now.

Gopnik also noted that many parents spend much energy trying to determine their children's future. However, parents can't give their children a good future, but they can give them a good past. That also applies to us. Can we as potential authors determine software engineering's future? Who knows? We can try. But it's not completely in our hands. And history teaches us that we have, on quite a few occasions, been wrong. For instance, Fred Brooks, in an excerpt from his book The Mythical Man-Month that appeared in IEEE Software, said, "[David] Parnas was right, and I was wrong [about information hiding] [12]." But can we give software engineering a good history? This is definitely much more under our control.

Defining IEEE Software's Future

Who controls the past ... controls the future: who controls the present controls the past. —George Orwell

Finally, one value of maintaining an accessible website about our history is being able to see how our past impacts IEEE Software's future. Orwell's quote from 1984 (don't forget that IEEE Software started in 1984) in many ways reflects the magazine's situation. The most obvious example is the impact factor—the frequency with which the average article or paper in a publication has been cited in particular years. Although the impact factor is based on past data, it directly influences a publication's reputation and future. A high impact factor normally attracts more high-quality contributions. High-quality articles and papers are normally cited more, which might further increase the impact factor. And vice versa: publications with a low impact factor normally attract fewer high-quality contributions, which might start a vicious cycle of decreasing impact factors.

Up to now, the IEEE Software history website has been an experiment. It's still a prototype, and we're still experimenting with different ways of presentation, adding new content and releasing changes frequently.

You can help contribute to this history. For example, write great new articles for IEEE Software. Invest the effort to find previous research and connect your research to it. Or, promote historical content in any medium—for example, by using that content in education and as inspiration in daily work.

Sidebar A: Other Histories of Software Engineering

The IEEE Software history website (obren.info/ieeesw) complements other resources describing software engineering history, such as these (links to which are also on the website):

- "History of Software Engineering"; en.wikipedia.org/wiki/History_of_software_engineering[a]
- N. Wirth, "A Brief History of Software Engineering," IEEE Annals of the History of Computing, vol. 30, no. 3, 2008, pp. 32–39.
- "A Brief History of Software Engineering," Viking Code School; www.vikingcodeschool.com/ software-engineering-basics/ a-brief-history-of-software-engineering.

- A Brennecke and R. Keil-Slawik, Eds., Position Papers for Dagstuhl Seminar 9635 on History of Software Engineering[b], 1996.

[a]https://en.wikipedia.org/wiki/History_of_software_engineering
[b]www.dagstuhl.de/Reports/96/9635.pdf

Sidebar B: IEEE Software Bibliometric Data

The IEEE Software history website (obren.info/ieeesw) combines data from the IEEE Computer Society Digital Library (CSDL), IEEE Xplore, and Google Scholar. The CSDL and Xplore provide useful data about the number of articles and authors.

Approximately 4,500 IEEE Software articles are indexed in Xplore. However, this number includes front and back covers, tables of contents, and ads. I built a script that extracts only articles with authors. This leaves around 3,250 "proper" articles. Approximately half of those articles are peer reviewed; the other half includes columns and invited content.

In total, more than 4,200 authors have contributed to IEEE Software. Of those authors, 819 have contributed multiple times—for example, Diomidis Spinellis (75 articles), Grady Booch (68), Robert Glass (57), Christof Ebert (46), and Forrest Shull (43). These 819 authors authored or coauthored approximately two-thirds of the articles. Fifty-one percent of the articles (mostly department articles) have one author; 49 percent have multiple authors.

Figures A and B shows the number of authors and articles

per year.

The history website also contains citation data extracted from Google Scholar in February 2017:

- The cumulative IEEE Software citation count is 161,042 (the sum of all "cited by" fields).
- The magazine's h-index is 181; approximately one-half of the citations are from these top 181 articles.
- The most cited year is 1990, followed closely by 2003 and 1994.
- The most cited articles are "The 4+1 View Model of Architecture" (2,786 citations), "Reverse Engineering and Design Recovery: A Taxonomy" (2,594 citations), and "Software Risk Management: Principles and Practices" (1,925 citations).

Figure C shows the number of IEEE citations per year of publication.

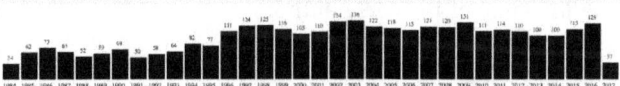

Figure A: The number of authors per year in IEEE Software.

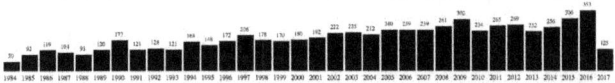

Figure B: The number of articles per year in IEEE Software.

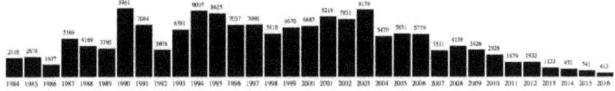

Figure C: The number of IEEE citations per year of publication for IEEE Software.

Sidebar C: History Tweets—Injecting the Past into Social Media

The IEEE Software history website (obren.info/ieeesw) also promotes IEEE Software on social media. From October 2016 to June 2017, to draw attention to the magazine's 200th issue (July/Aug. 2017), I've been daily tweeting IEEE Software covers and quotes and interesting historical findings. In this way, each IEEE Software issue has been mentioned at least once before the publication of our 200th issue.

The tweets have provided an interesting way to engage with a broader, younger audience. Many of the old articles from the 1980s have received significant attention. Social-media interaction has also enabled us to reconnect with some of the early authors.

Also, on Twitter under #SE_history (twitter.com/hashtag/se_history) are more than 500 tweets about IEEE Software history. We plan to tweet there again as new issues are added. Figure D shows a few interesting tweets.

Figure D: Some interesting tweets from the IEEE Software history website:

#SE_History @ieeesoftware Jan 1984 review of @Grady_Booch's book and his crazy idea of "object-oriented design".
obren.info/ieeesw/

Software Engineering with Ada—Grady Booch (The Benjamin/Cummings Publishing Company, Inc., Menlo Park, Calif., 1983, 491 pp., $20.35)

The Ada programming language has finally, in 1983, achieved not perfection, but ANSI standardization with almost-production-level compilers that are not bug-free, but validated. Ada is Pascal "grown up." We now have a tool that embodies modern software engineering principles and supports computer programming as a human activity. Its

......

Booch offers a software design methodology, which he calls "object-oriented design" in contrast to earlier popular methods he designates as either functional or data-oriented. In contrast to suggestions that we either identify a program's principal function and describe it in the top box in a hierarchy chart or carefully identify the patterns of data and their flows, Booch suggests, "Define the problem; develop an informal strategy; formalize the strategy." This is like classical problem solving, and it, like Ada, imposes no (or comparatively few) constraints on the early stages of the process.

IEEE Software, January 1984, page 119

4:44 PM - 9 Nov 2016

12 Retweets 19 Likes

Insights from the Past

The story of C++ >> CPP::CPP(Programmer *p) { if (p->serious) p->enjoyment++; }
@ieeesoftware **Jan 1986** #SE_history
obren.info/ieeesw/

"The **C++ programming language** was designed to make the task of programming **more enjoyable for the serious programmer.**"

Bjarne Stroustrup, IEEE Software, **January 1986**, p. 71

1:01 PM - 21 Nov 2016

13 Retweets **11** Likes

The first computer bug - taped to the the first bug tracking system, #SE_history @ieeesoftware **May 1987** obren.info/ieeesw/?quotes ...

"[T]he term [debugging] was first applied to **a hardware bug - a moth** in the circuitry of Mark II" (R.E. Seviora, Knowledge-Based Program Debugging Systems, IEEE Software 1987, no. 3, p. 20)

A computer log entry from the Mark II, with a moth taped to the page

10:59 AM - 30 Nov 2016

18 Retweets **15** Likes

#Architecture is not about #software but about #people.
#Code doesn't care about how cohesive or decoupled it is.

@jcoplien #SE_history '99

"*[A]rchitecture* is not so much *about the software*, but about the people who write the software. The *core principles* of architecture, such as coupling and cohesion, *aren't about the code*. The *code doesn't 'care'* about how cohesive or decoupled it *is*; if anything, tightly coupled software lacks some of the performance snags found in more modular systems. But *people do care* about their *coupling to other team members*." (James O. Coplien, Guest Editor's Introduction: Reevaluating the Architectural Metaphor-Toward Piecemeal Growth, IEEE Software 1999, no. 5, p. 40)

8:49 AM - 24 Feb 2017

16 Retweets 26 Likes

"The best way I know to #deliver sooner is to do less."

J. B. Rainsberger @jbrains
#SE_history @ieeesoftware 2007

"*Experienced programmers plan*, while *junior programmers jump into* their work. Some simpler *personal planning techniques* can help you eliminate waste when you work, write less code, design more simply, inject fewer defects, and generally *deliver sooner*. ... [T]he best way I know to *deliver sooner* is to *do less*." (J.B. Rainsberger, Personal Planning, IEEE Software 2007, no. 1, p. 16)

4:57 PM - 17 Apr 2017

42 Retweets 43 Likes

 42   ♡ 43

References

1. R.L. Glass, "'Silver Bullet' Milestones in Software History," Comm. ACM, vol. 48, no. 8, 2005, pp. 15–18.
2. M. Fowler, "Semantic Diffusion," Dec.2006; martinfowler.com/bliki/SemanticDiffusion.html[4].
3. K. Beck et al., "Manifesto for Agile Software Development," 2001; agilemanifesto.org.
4. B.D. Shriver, "From the Editor-in-Chief," IEEE Software, vol. 1, no. 1, 1984, pp. 4–5.
5. J.O. Coplien, "Reevaluating the Architectural Metaphor: Toward Piecemeal Growth," IEEE Software, vol. 16, no. 5, 1999, pp. 40–44.
6. P. Hsia, "Learning to Put Lessons into Practice," IEEE Software, vol. 10, no. 5, 1993, pp. 14–17.
7. P. Wegner, "Capital-Intensive Software Technology," IEEE Software, vol. 1, no. 3, 1984, pp. 7–45.
8. S.-K. Chang, "Visual Languages: A Tutorial and Survey," IEEE Software, vol. 4, no. 1, 1987, pp. 29–39.
9. "Inventing the Future" (interview with A. Kay), IEEE Software, vol. 15, no. 2, 1998, pp. 22–23.
10. C. Alexander, "The Origins of Pattern Theory: The Future of the Theory, and the Generation of a Living World," IEEE Software, vol. 16, no. 5, 1999, pp. 71–82.
11. A Gopnik, The Philosophical Baby, Farrar, Straus and Giroux, 2009.
12. F. Brooks, "The Mythical Man-Month after 20 Years" (book excerpt), IEEE Software, vol. 12, no. 5, 1995, pp. 57–60.

---

[4]https://martinfowler.com/bliki/SemanticDiffusion.html

6. The Researchers-Practitioners Manifesto

My goal is to stimulate practitioners to start doing research in practice and to organizes themselves OUTSIDE current academic structures. I believe that the only way to bridge the practice-research gap is for practitioners to start doing research and to bottom-up build necessary organizations and structures.

6.1 As Individuals

We recognize that being a researcher is a call rather than a job title.

Research can be conducted in many settings and outside research institutions. Research is conducted by motivated and competent individuals and groups, not necessarily working in research institutions, or R&D departments.

We investigate realistic real-world settings.

We conduct our research in a messy world of practice, in realistic settings, with real users.

We view research as primarily being about knowledge and secondary about research methods.

Research methods are a means to obtain knowledge. Knowledge is power.

We choose research methods based on the questions that need to be answered, not the other way around.

No method is universally good or bad. It may be more or less appropriate for a research question.

We prioritize our research questions according to their relevance and importance.

We avoid choosing research questions according to how well they fit our preferred research methods. We value knowledge, impact, and relevance more than obtaining grants, patents, or publications.

We value knowledge over innovation.

We do research to obtain knowledge. We reuse existing knowledge as much as possible. We avoid re-inventing the wheel. We always connect our research to the existing body of knowledge. We know our field and its history. Our work is always standing on the shoulders of the giants. We make this fact explicit.

We share our results and make our claims explicit, public, and open to critical reflection and discussion.

We transparently report research results. Our methods, motivation, funding, and conflicts of interests are always explicitly stated.

6.2 As a Community

We recognize the importance of belonging to and building a community of peer researchers-practitioners.

We organize ourselves in communities to share knowledge, peer-review each other work, and connect and make easily accessible our work.

We recognize the importance of peer-review.

It is easy to get carried away when doing creative and exciting work. We need constructive feedback to avoid making errors. Research contribution is not research contributions if it has not been peer reviewed. We pro-actively seek peer review of our work. We pay back by peer reviewing others work.

We value collaboration and results over belonging to a formal institution.

We recognize that the community is a group of researchers collaborating to make each other's research better, not necessarily a formal organization.

II DESIGN & RESEARCH

7. Design-Based Research

The basis for most of my research is design. I build experimental systems to explore new domains. This essay discusses why design can be a basis for doing research, focusing on what we can learn when we engage in design of computer systems.[1]

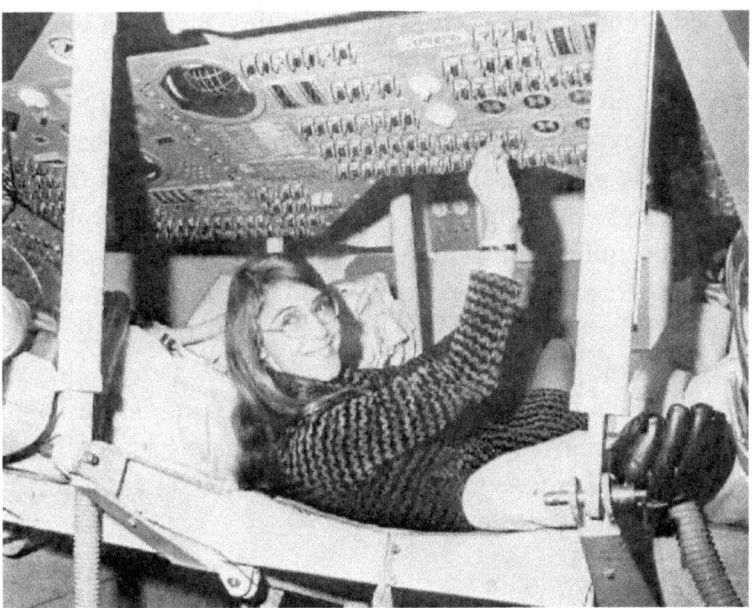

Margaret Hamilton during her time as lead Apollo flight software designer. Wikimedia Commons

---

[1]This chapter is based on the article Design-based research: what we learn when we engage in design of interactive systems, interactions 18, 5 (September 2011), 56-59.

Introduction

More than 20 years ago, Fred Brooks asked, "*Is interface design itself an area of research, producing generalizable results?*" [1]. He elaborated that a major issue that puzzles the human-computer interaction community is the tension between narrow truths proved convincingly by statistically sound experiments, and broad truths, generally applicable, but supported only by possibly unrepresentative observations—that is, results indisputably true but disputably applicable, and results indisputably applicable but perhaps overly generalized.

Brooks' question is still relevant. In this chapter, I explore the view that the design of complex and novel interactive systems can itself be an area of research, complementing other forms of research, and that it is capable of producing useful and trustworthy results. I call this form of research design-based research, a method of inquiry aimed at exploiting the opportunities that designing complex interactive systems provides to advance our understanding of the problem we are solving, the process we are following, and the solution we are building. While many designers and researchers already conduct this form of research, and the idea of design-based research is not new (for example, [2]), there have not been many attempts to explicitly define this method and address the following questions:

- What can we learn when we engage in the design of interactive systems?
- What kind of generalizable knowledge can we get from design?
- What is the relationship of design-based research to theoretical and experimental methods?

- Why can design reveal things that other research methods cannot?
- What makes design-based research trustworthy?

What We Learn When We Engage in Design

Design can be described as a sequence of decisions made to balance design goals and constraints. In any design activity, designers make a number of decisions, trying to answer the following questions [3]:

- How will the design process advance?
- What needs and opportunities will the design address?
- What form will the resulting product take?

These decisions must be made in every design effort, although they may not be explicit, conscious, or formally represented. In routine design, these decisions are straightforward, requiring little learning by designers. In challenging or innovative designs, however, these decisions can be complex and interdependent, requiring extensive investigation, experimentation, and iterative improvement. In such situations, designers may acquire important new understandings. This ability to acquire new knowledge through design provides the basis for doing research, which aims at capturing this new knowledge and making it available to a broader audience. We may group lessons that we learn in design into three categories: design procedures, problem analysis, and design solutions.

A design procedure specifies which processes and individuals are involved in a design. Designers often have to develop specialized procedures to respond to a specific design challenge or the context in which the design is being constructed.

Problem analysis describes our current understanding of the problem we are facing. One of the characteristics of design is that we never start with a clear understanding of the problem, and one of the chief services of a designer is helping clients to discover what they want designed [4]. Design problems are often full of uncertainties about both the objectives and their priorities, which are likely to change as the solution implications begin to emerge. Problem understanding evolves in parallel with the problem solution, and many components of the design problem cannot be expected to emerge until some attempt has been made at generating solutions [5]. Simon, in what he calls "designing without final goals," wrote that a goal of design may actually be understanding the problem and generating new goals, elaborating that the idea of final goals and a static problem definition is inconsistent with our limited ability to foretell or determine the future [6].

A *design solution* describes the resulting product, the outcome of designers' efforts to address challenges, satisfy constraints, exploit opportunities, and balance the trade-offs identified in the problem analysis. The design solution evolves over the design process as designers deepen their understanding about the design context and problem.

Generalizable Knowledge

To be regarded as a research contribution, design activity should go beyond simply refining practice and also address theoretical questions and issues. Design-based research extends ordinary design activity with a goal of developing generalizable knowledge. In a normal design effort, the primary goal is to create a successful product, and lessons learned are restricted to the particular design and the people involved in it. In the process of generalizing, however, a designer-researcher expands his focus beyond the current design situation, viewing the design problem, solutions, and procedures as instances of more general classes. For each of the collections of lessons learned, we may identify the corresponding type of generalization: domain theories, design frameworks, and design methodologies [3].

A domain theory is the generalization of a problem analysis. A domain theory might be about users of interactive systems and how they learn to use and interact with the systems, or about the context of the system usage and how it influences the user and interaction. A domain theory is a means of understanding the world, not the design solution or procedure.

A design framework is a generalization of the design solution. Design frameworks describe the characteristics that a design solution should have to achieve a particular set of goals in a particular context. In other words, a design framework represents a collection of coherent design guidelines for a particular class of design. Design patterns and software architectures are prominent examples of this class of generalization.

A design methodology is a generalization of a design procedure. In contrast to design frameworks, a design methodology provides guidelines for the design process rather than the product. In general, a design methodology describes a process for producing a class of design solutions, the types of expertise required, and the roles of people with these types of expertise.

In general, design-based research cannot develop "grand" or universal theories and frameworks. Rather, it develops generalizable knowledge with an intermediate theoretical scope, covering a gap between a narrow explanation of a specific design and a broad, more generic account that does not limit the design to a particular situation.

Design-Based Research and Other Research Methods

Generalized knowledge can also be derived by using other empirical or theoretical research methods. Design-based research, however, can produce knowledge that normally could not be generated by isolated analysis or traditional empirical approaches, and therefore complements existing empirical and theoretical research methods. Design-based research facilitates disciplined, systematic inquiry into a real-world context while simultaneously doing justice to its complexity. It is conducted in messy, but entirely realistic, situations and while it produces claims with less certainty and replicability than other research methods, it can extend our area of inquiry beyond the scope of these methods.

Design-Based Research Versus Other Empirical Research Methods

Controlled experiments are one of the most powerful and conclusive forms of empirical research, used to establish the relationship between the cause and the effect by manipulating an independent variable to see how it affects a dependent variable. Although conducting experimental research has enormous benefits, it also has some serious limitations. The controlled experiment may be conducted only if we know the relevant variables involved in research, we can define important relationships among the variables, and we can control all extraneous variables that might affect the outcome. These conditions significantly limit the scope of experimental research, and in many real-world situations we cannot fulfill them, as a researcher usually cannot maintain control over all factors that may influence the result of an experiment. Attempting to simplify a real-world situation so it can be subjected to experimental research often leads to studying unrealistically simple situations. Attempting to establish an experimental control in a real-world setting, on the other hand, may lead to negative phenomena, such as the Hawthorne effect, in which those who perceive themselves as members of the experimental, or otherwise favored, group tend to outperform their controls, often regardless of the intervention.

Design-based research can produce knowledge that normally could not be generated by theoretical analysis or traditional empirical approaches.

Although significantly different, controlled experiments and design-based research are compatible forms of research that can be and often are used together. Controlled experiments, for example, can guide design decisions and test particular elements of a design on a smaller scale and in more controlled conditions. A decision about which input control to use in a user interface, for instance, may be based on the results of a controlled experiment comparing the efficiency of users' data input with several alternatives. Controlled experiments may provide reliable information that something "worked," but they often do not provide sufficient information about exactly what it was that worked, or why or how it worked. Design-based research can help us to characterize and identify relevant variables, create an explanatory framework for the results of the experiments, and provide us with more insights about why and how some elements of a design work.

Ethnographic research and field studies attempt to characterize relationships and events that occur in some setting to produce rich descriptions that make it possible to understand what is happening and why. In contrast to design-based research, however, there is no attempt to change this situation. Design-based research complements these methods by enabling us to learn more about the real world by changing it and reflecting on our experiences in understanding problems, design solutions, and procedures.

Design-Based Research Versus Theoretical Research

Design-based research requires an alternative view on the relationship between theory and practice in which neither is taken as primary. Design activity is often driven by existing theories, and at the same time it can provide a constructive environment for theory development. Design process can often reveal theoretical inconsistencies more effectively than analytical processes, while designing a concrete system based on some theory requires that it be fully specified [4]. On the other hand, the development of theoretical constructs and standards without their grounding in a concrete design often leads to a range of problems, as shown in Henning's discussion about the reasons for the decline of CORBA (Common Object Request Broker Architecture) [7]. Henning concluded that standards consortia must ensure they standardize only existing best practices and that no standard should be approved without a reference implementation and without having been used to implement a few projects of realistic complexity.

Why Design Can Reveal Things That Other Methods Cannot

Design-based research complements existing research methods in its ability to employ in a greater amount the tacit, implicit, intuitive knowledge and skills of both designers and users. Schön calls such knowledge knowing-in-action, revealed only in the way in which we carry out tasks and approach problems: "The knowing is in the action. It is revealed by the skillful execution of the performance—we are characteristically unable to make it verbally explicit" [8].

In other words, though we cannot explain such knowledge and skills, we can demonstrate them by being engaged in a particular activity. This observation is supported by studies of embodied cognition, which emphasizes the formative role that the environment plays in the development of cognitive processes [9].

A design activity can set in motion our intuitive and tacit knowledge accumulated through years of research and experience. Much of such valuable knowledge is not captured in existing theories and guidelines. Often, we are not aware that we possess it. Glass, for example, noted that actions of designers are often implicit and intuitive, defining intuition as *"a function of our mind that allows it to access a rich fund of historically gleaned information we are not necessarily aware we possess, by a method we do not understand"* [10].

Glass further elaborated that our unawareness of such knowledge does not mean we cannot use it. Designers, for example, often cannot explain their own creative processes, but, through design, they can apply and materialize these creative skills in solving a range of complex problems.

Our intuition and tacit skills also play an important role in understanding and setting problems from messy and ill-defined situations. By engaging in design, we can better understand real-world, ill-defined, and wicked problems (as discussed earlier). Similarly, through design we can better understand users' needs, as our users often cannot precisely explain to us what they want unless we present them with some version of a design solution [4]. Moreover, by engaging users in design, we may employ their knowledge about their domains, as well as their creativity.

While design itself adds discipline and professional attitude to tacit, implicit, and intuitive knowledge and skills, design-based research may be viewed as an attempt to increase awareness of such knowledge and to support, capture, generalize, and share this knowledge beyond the design community. Therefore, design-based research can be an especially valuable method of inquiry in domains such as interaction design, which does not have strong theories, models, and laws to conduct extensive theoretical analyses, simulations, or experiments, but does have practitioners and users who have some (often tacit, implicit, intuitive) knowledge and skills related to the domain.

Trustworthiness of Design-Based Research

While design-based research puts trust in designers' skills, ingenuity, and ability to correctly observe and generalize issues observed in a design process, this trust should not be blind. Results of research must be presented in a way that enables readers to clearly understand the motivation and reasoning behind particular claims. This means that designers must provide sufficient information so that generalized claims can be verified. Trustworthiness of design-based research comes from making the reasoning behind generalized claims explicit, public, and open to critical reflection and discussion.

Conclusion

The study of interactive systems requires the selection of appropriate methods from a wide array for each research question asked. For many of these questions, theoretical analyses, controlled experiments, or ethnographical research are the best methods. However, design-based research can produce knowledge that normally could not be generated by theoretical analysis or traditional empirical approaches. It can help us to better understand the problem and ask better research questions, often having a pioneering role in settling a new research territory that can then be "occupied" by other research methods.

If lessons learned in design are to become accepted as serious scholarly endeavors within and outside our discipline, we need to take responsibility for creating standards that make such research recognizable and accessible to other researchers. In particular, more work is necessary to create a framework that can enable us to combine results of design-based research with results from other forms of research. And we as a community need to better understand the scope and limitations of design-based research to be able to critically review contributions of this kind.

References

1. Brooks, F.P. Grasping reality through illusion-interactive graphics serving science. Proc. CHI '88. ACM, New York, 1988, 1–11.
2. Zimmerman, J., Forlizzi, J., and Evenson, S. Research through design as a method for interaction design research in HCI. Proc. CHI '07. ACM, New York, 2007, 493–502.
3. Edelson, D.C. Design research: What we learn when we engage in design. Journal of the Learning Sciences 11, 1 (2002), 105–121.
4. Brooks, F.P. The Design of Design: Essays from a Computer Scientist. Addison-Wesley Professional Boston, MA, 2010.
5. Lawson, B. How Designers Think: The Design Process Demystified, 4th Edition. Architectural press, Oxford, UK, 2005.
6. Simon, H.A. The Sciences of the Artificial, 3rd Edition. MIT Press, Cambridge, MA,1996.
7. Henning, M. The rise and fall of CORBA. Comm. of the ACM 51, 8 (Aug. 2008), 52–57
8. Schön, D. The Reflective Practitioner. Basic Books, New York, 1983.
9. Dourish, P. Where the Action Is: The Foundations of Embodied Interaction. MIT press, Cambridge, MA, 2001.
10. Glass, R.L. Software Creativity 2.0, developer.* Books, Atlanta, GA, 2006.

8. Doing Design-Based Research in Practice

The previous chapter presents the general description of design-based research. This essay presents more practical advice about how to do design-based research in practice.[1]

System/360 Model 91 Panel at the Goddard Space Flight Center (the image was taken by NASA sometime in the late 60s). Fred Brooks managed the development of IBM's System/360 family of computers and the OS/360 software support package. He later wrote about the lessons learned in his seminal book The Mythical Man-Month.

---

[1]This chapter is based on the article Doing research in practice: some lessons learned, XRDS 20, 4 (June 2014), 15-17.

In the previous chapter I argued that, due to extensive tacit and intuitive knowledge and skills, practitioners may be able to acquire unique understandings. In this chapter I want to argue that the knowledge you acquired through practice is of limited value if your learning process is not disciplined. Our experiences differ a lot, and our abilities to communicate and understand it are also different. Practice-based research is very limited if it is not accompanied with the following features: prepared mind, systematic documentation, generalization, evaluation and iterations [1].

Prepared Mind. Louis Pasteur famously noted, "*chance only favors the mind which is prepared.*" Pasteur was speaking of Danish physicist Oersted and the almost "accidental" way in which he discovered the basic principles of electro-magnetism. He elaborated that it is not during accidental moments that an actual discovery occurs: The scientist must be able, with prepared mind, to interpret the accidental observation and situate the new phenomena within the existing work. Similarly, our learning in practice is limited without awareness about the context and existing solutions. We may build the wrong solution, or waste our time by "reinventing the wheel," instead of exploiting existing work, available, theories and empirical results. Brooks similarly agued that practitioners should know exemplars of their craft, their strengths and weaknesses, concluding that originality is no excuse for ignorance [2]. Furthermore, if we are not able to connect our observations to a broader context and existing work we may not be able to judge the relevance and importance of our observations and we may miss the opportunity to make some important discovery.

Systematic Documentation. To support research, your practical experiences should be documented. We keep forgetting things, and our memory changes over time. Keeping systematic documentation also enables retrospective analyses and discovery of new findings even after our projects finish. It is important to document all-important decisions, describing the limitations and failings of the design, as well as the successes, both in implementation and usage. Of particular value is documenting the rationale for our decisions, describing why we did something, not only what we have done. This is not easy, as it often requires making explicit the elements practitioners use intuitively. But such documentation is extremely valuable, especially for others, to understand why some things worked well or not.

Sharing and Generalization. In a normal design effort the primarily goal is to create a successful product, and lessons learned are restricted to the particular design and the people involved in it. To be useful to others, some effort has to be invested in generalizing these lessons. Generalization enables correlating different experiences, which otherwise may look too specific.

Knowledge obtained from practice can be viewed as a generalization of experiences. In a normal design effort the primarily goal is to create a successful product, and lessons learned are restricted to the particular design and the people involved in it. To be useful to others, some effort has to be invested in generalizing these lessons.

Generalization enables correlating different experiences, which otherwise may look too specific. In the process of generalization, practitioners need to expand their focus beyond the current design situation, viewing the design problem, solutions, and processes as instances of more general classes. For each of the collections of lessons learned as discussed in the previous section, we may identify the corresponding type of generalizations (see the chapter Design-Based Research): domain theories as generalizations of problem analysis, design frameworks or patterns as generalizations of design solutions, and design methodologies as generalizations of design procedures.

However it is also important not to "over-generalize." Practice usually cannot provide us with insights to develop "grand" and universal theories. Rather, emphasis should be between narrow truths specific to some situation, and broader knowledge covering several similar situations.

Evaluation. Evaluation is a necessary part of any learning process. This is especially true for domains, such as software engineering or human-computer interaction, where we have to deal with complex human and social issues for which we do not have strong theories, models and laws. By evaluation I do not necessarily mean formal evaluation activities conducted in "laboratories" (even thought these may be used sometimes). I see evaluation as a systematic effort to get feedback on our findings and quality of our work that is not based on our intuition. Even simple techniques, such as peer reviews may be incredibly efficient in identifying shortfalls in our problem analysis, the solution, and the design procedure. In order to enable such a process, practitioners need to be prepared to make their reasoning explicit, public, and open to critical reflection and discussion. The key is to make your intuitive decisions more explicit and "vulnerable" to the critique of others and empirical findings.

Iterations. Lastly, to maximize learning, all of the previous elements should be applied in a number of iteration. While a single event can have an impact, it usually takes many events to extract general features and generate rules from experience. In everyday work, you should try to combine elements of preparation, actual practical work, and evaluation (see Figure 1a).

At the begin of each learning cycle you should spend some time preparing for you actions, reflecting on previous actions and hypothesize about how you expect that your new actions will affect outcomes. I call this phase a "thought experiment." In this phase, it is important to make explicit practitioners' expectations and assumptions, so others can understand and evaluate them. After actual practical work, there should be an evaluation where outcomes of your actions are reviewed in the light of their original assumptions. Results of evaluations can further improve your understanding, and serve as a basis for new cycles. Ideally, this process should also include a preparatory research, where practitioners collect as much information about existing work as possible at the beginning, and make their goals explicit and clear. This model may be too idealistic for practical work. However, I find this model a very useful guide to structure and add some discipline to everyday work, and in that way maximize learning from practice and obtain knowledge that may be interesting for the research community.

Learning should not stop at the end of a project. New insights and broader generalizations will often occur through retrospective analyses of lessons learned and data collected through a whole project. Brooks, for example, spent several years analyzing and reflecting on the lessons learned in design of OS/360, producing the influential The Mythical Man-Month [3]. These macro-iterations of retrospective analyses can also happen on a broader scale, covering several projects from different contributors. For example, in special issues of journals, editors often spend some time summarizing and generalizing findings from individual articles.

(a) mini iterations of practical work, evaluation / critique and thought experiments

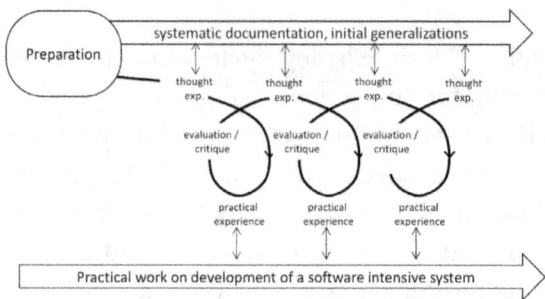

(b) macro-iterations of practice-base research including retrospective analyses

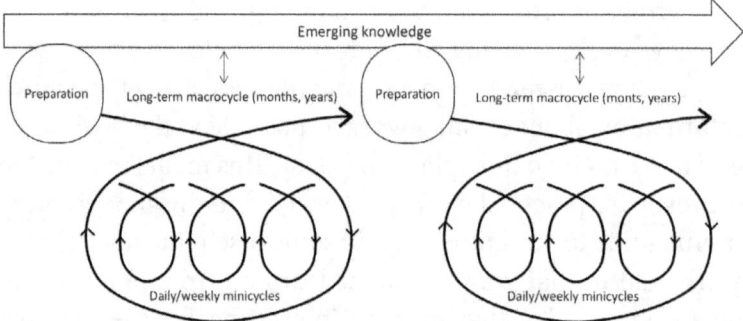

Figure 1: A simple model of iterations in practice-based research. We can view a practice-based research process as composed of mini-iterations of preparations, practical work, and evaluations, occurring continuously through the project (a), embedded in long-term macro-iterations that include research preparation and retrospective analyses and reflections (b), covering the period of one or more projects. Macro-iterations can itself be connected so that results of one research project guide preparations of another research project.

References

1. Edelson, D.C. Design research: What we learn when we engage in design. Journal of the Learning Sciences 11, 1 (2002), 105–121.
2. Brooks, F.P. The Design of Design: Essays from a Computer Scientist. Addison-Wesley Professional, 2010.
3. Brooks, F.P. The Mythical Man-Month. Addison-Wesley Professional, 1995.

9. Design As a Political Activity

The following essay does not directly discuss the research in practice. However, it addresses one factor that will inevitably affect any research and design activates in practice - politics. Everything that you do in practice will have an immediate impact on people and their behavior, and lead to a number of complex political situations. I wrote this essay as a reaction to the fact that, while there are many articles about politics in design, none of them define the meaning of the term 'politics'.[1]

Victoria Woodhull a first female candidate for U.S. president addressing congress in 1872.

---

[1] This chapter is based on the article Design as a political activity: borrowing from classical political theories, interactions 22, 6 (October 2015), 42-45.

Introduction

Several design scholars have suggested that design is a political activity. Jonas Löwgren and Erik Stolterman, for instance, claimed that all designs are manifestations of political and ideological ideas because design outcomes influence our lives [1]. Björn Franke argued that a design is a political decision about how people should live, communicate, or behave (see www.designaspolitics.com). Franke also maintained that we could view politics as a form of design because it involves planning, making decisions, and creating laws. And Michael Bierut argued in similar terms about graphic design: *"Much, if not most, graphic design is about communicating messages, and many of these messages are intended to persuade. This places its practice clearly in the realm of politics"* [2].

Discussions about the political aspects of design are not new; however, most of these discussions have been vague. Calling design a political activity because it influences the lives of people does not say much—almost everything we do directly or indirectly influences others. And calling politics a form of design does not help clarify the distinctions between these two terms. With such broad and vague discussions, we risk inflating the already loaded terms design and politics, making constructive discussions about them difficult, if not impossible.

With this article I want to support ongoing discussions about the political aspects of design by exploring the meaning of the word politics more deeply. I base my exploration on James Alexander's recent analysis of definitions of politics in five classical political theories [3]. I will argue that classical political theories provide strong, concrete support for claims that design is a political activity. Moreover, looking at design through the lenses of classical political theories reveals interesting and complex sets of political situations.

Classical Political Theories

Politics is a widely used and loaded term. Alexander found that many influential political theories contradict one another in their definitions of politics [3]. He noted, however, that these theories do have several common characteristics and that each theory sheds a different light on the meaning of the word. Here, I review these common characteristics and briefly present five classical political theories that Alexander attempted to generalize.

A central theme in definitions of politics is a rule, a common standard that regulates some aspect of human lives. Politics is normally defined as an activity through which people make, preserve, and change the general rules under which they live [3].

Political theorists do not view politics as being identical to rules or as a simple execution of the rules. Rather, politics is an activity related to the introduction or changes of rules in situations where there is a division of people in two classes:

- the rulers, a group of people who have (formal or informal) powers to define the rules that other people need to follow, and
- the ruled, a group of people who are (formally or informally) following the rules defined by rulers.

Political theorists generally agree that in politics the ruled are not simply passive followers of rules; the ruled have the opportunity to influence, actively change, or disrupt the rules.

Robin George Collingwood defined politics as the science of *"rightness or conformity to rule"* [4]. According to Collingwood, political action is *"the making and obeying the laws ... regulation, control, the imposition of order and regularity upon things"* [4]. Through politics, rulers and ruled are attempting to meet shared standards so that order can be achieved. For Collingwood, politics is related to any activity associated with rules (the view that I share in this article):

The rules of any corporation, the statutes of a company, the regulations of a club, the routine of a family, are all political facts, and no less political are the rules which a man makes for his own guidance, and revises [them] from time to time as occasion demands [4].

To be political, Collingwood stated, an activity has to satisfy three "laws of politics" [5]:

- People involved in an activity are divided into a ruling class and a ruled class.
- The barrier between the two classes is permeable in an upward sense (i.e., the ruled can become the rulers).
- There is a correspondence between the rulers and ruled.

Michael Oakeshott argued that politics happens when persons without authority (i.e., the ruled) can approve or disapprove of rules, or offer their opinions about the need to change or not change these rules [6]. In Oakeshott's view, in a political activity the ruled respond to attempts by the rulers to lay down common standards. The rulers propose common standards, and the ruled can express judgments about how these standards affect their interests.

Hannah Arendt viewed persuasion as one of the most important elements of politics. To be political, says Arendt, means that everything is decided *"through words and persuasion and not through force and violence"* [7]. In political situations, *"men in their freedom can interact with one another ... as equals among equals ... managing all their affairs by speaking with and persuading one another"* [8]. In Arendt's view, to command (i.e., to rule) rather than to persuade involves pre-political ways of dealing with people, characteristic of life outside the political system.

Carl Schmitt maintained that a central element of a political activity is a decision [9]. Simply talking about rules does not make an activity political. An activity may be called political only to the extent that it crystallizes in a decision. According to Schmitt, in any political activity the rulers and ruled face the political imperative that a decision be made. How this decision is made is not a defining characteristic. Rather, what makes some activity political is the mere imperative that a decision is made. Schmitt also views politics in more negative terms, as nothing more than party politics, where people take sides to reach desired decisions.

Last, Jacques Rancière sees politics as a revolt against the ruling according to established rules [10]. In his view, ruling is the activity of police, while politics is anything that disrupts this activity. Rancière defines police order as a set of implicit rules and conventions that determine the distribution of roles in a community. A police order defines rules and imposes constraints on what can be thought, made, or done in a particular context. Politics, in Rancière's view, is an activity that challenges such police order and its rules.

Here, I only briefly sketched definitions of politics from classical political theories. For readers who want to explore this topic further, I recommend reading the work of the political theorists mentioned. I also recommend James Alexander's article for a nice overview [3].

Design as a Political Activity

Classical political theories provide strong, concrete support for claims that design is a political activity. Politics revolves around the process of defining rules and common standards that regulate human activities, and design is always about defining some such rules and common standards. Löwgren and Stolterman, for example, claimed that every design constrains our space of possible actions by promoting the usage of certain skills and focusing on the creation of certain outcomes [1]. Similar to the Rancière's view on the police order, each design imposes constraints on what can be thought, made, or done in a certain context.

What political theories add to current discussions about design and politics is the view that the mere existence of rules does make some activity political. Politics "happens" when rules can be introduced or changed based on the interaction between the rulers and the ruled. Consequently, we may say that design is a political activity because (and only when) design stakeholders can influence introduction and changes of design-related rules. Design professionals who do not allow others to influence such rules risk being perceived as dictators, as discussed in Alex Cabal's blog post "The Cult Of Design Dictatorship" (https://alexcabal.com/the-cult-of-design-dictatorship/).

Looking at design through the lenses of classical political theories reveals an interesting and complex set of political situations. One group of these situations is the interaction between design professionals and users. Different approaches to design promote different balances of power between design professionals and users. In user-centered design (UCD), for example, design professionals try to optimize a product around how users can, want, or need to use the product. Here, design professionals have most of the decision-making power, but users (or their representatives) are encouraged to provide feedback. Users can express their opinions and comments on proposed or implemented rules (e.g., through participatory design sessions, prototype and usability testing, or satisfactory surveys). In political terms, the rulers (design professionals) and the ruled (users) are attempting to agree on shared standards so that order can be achieved (as Collingwood suggests). Alternatively, we may view UCD as a political situation in which the rulers (design professionals) are attempting to define common standards and the ruled (users) respond to these attempts by making judgments about how these standards affect them (as Oakeshott suggests).

Co-design and participatory design promote more equal relations between design professionals and users (e.g., [11]). In these approaches, the goal of design professionals is not to make final design decisions. Rather, these approaches are empowering, encouraging, and guiding users to make these decisions for themselves. In political terms, these initiatives attempt to blur the distinction between the rulers and the ruled, making them "equals among equals," as Arendt suggests.

We may also talk about "mass politics" as a form of interaction between design professionals and users. Various consumer product "revolutions" may be viewed as such. Users have often rejected existing products and begun using new ones perceived as better or more desirable (e.g., gesture-based vs. keyboard-based smartphones). Here, individual users do not have a direct influence on design decisions. However, they can accept or reject competing products through their free choice in the market. In that way, they can indirectly stimulate companies to react and change their designs. In political terms, these situations may be viewed as revolts against established rules (as Rancière suggests). Such "mass politics" can also force other changes in design companies. In 2014, for example, a number of users boycotted the Firefox Web browser because of the CEO's stance on gay rights. This boycott significantly contributed to the pressure that led to the CEO's resignation.

Another set of political situations in design relates to the rules that regulate a design process. Design professionals and other stakeholders need to agree on a number of rules that coordinate their work. Defining a design process, deciding on a budget, setting priorities, negotiating deadlines, and selecting tools and materials are complex political activities with many stakeholders. Interactions between design professionals and clients are one example. On the one hand, clients define the terms and conditions of contracts and are responsible for providing financial and other support. On the other hand, clients depend on the expertise and ideas of design professionals. Design professionals are not mere executors of the client's wishes, and they are expected to be innovative. But to get their ideas accepted, design professionals need to interact with clients and persuade them of the rightness of particular design choices. This aspect is nicely illustrated in an anecdote provided by Herbert Simon. Simon had asked Ludwig Mies van der Rohe how he got a client to sign off on a house that was radical for 1930. Mies apparently replied, "*He wasn't happy at first. But then we smoked some good cigars ... and we drank some glasses of a good Rhine wine ... and then he began to like it very much."*

This anecdote may be described as a political situation viewed in Arendt's terms. Decisions are reached "*through words and persuasion,*" and people manage their affairs "*by speaking with and persuading one another.*"

In some cases, complex political interaction among design stakeholders may lead to negative consequences and "party politics." A typical example is "design by committee." Fred Brooks argued that outcomes of a design by committee lack focus and result in impractical products with too broad functionality. Brooks elaborated that the people in committees, in order to protect their own interests, are often reluctant to reject any request:

Each player has a wish list garnered from his constituents and weighted by his personal experiences. Each has both an ego and a reputation that depend on how well he gets his list adopted. Logrolling is endemic—an inevitable consequence of the incentive structure. "*I won't naysay your wish, if you won't naysay mine*" [12].

The need of design professionals and other stakeholders to make decisions within limited timeframes further emphasizes the political aspect of design. All projects have deadlines. Design professionals often need to make a number of agreements and compromises to meet these deadlines. Even when a strict deadline is not imposed, the dynamics of the design process may put pressure on design professionals to make decisions quickly. Bryan Lawson, for example, noted that procrastination as a strategy in design is deeply flawed [13]. He elaborated that once a design problem has been identified, it is no longer possible to avoid making decisions about a design outcome: "*In many real-life design situations it is actually not possible to take no action. The very process of avoiding or delaying a decision has an effect!"* [13]. For example, if a new road is planned but the route remains under debate for any lengthy period, the property in the region will likely change in value. Here we have a typical Schmitt's situation where the rulers and ruled face the political imperative that a decision be made.

Conclusion

Examples in the previous section are just some of the possible political situations in design. Design professionals may also be involved in other political situations, including:

- politics surrounding public policies, as illustrated by the long-running Interactions forum of the same name,
- workplace politics of organizations in which design professionals operate,
- politics of design educational institutions and funding agencies, and
- politics of professional organizations (such as ACM).

And the list could be extended even more. My goal is not to provide an elaborate rundown of all possible political situations in design. Rather, I want to illustrate that the space of political situations in design is broad and diverse.

Consequently, we need to be thoughtful about the political aspects of design. Politics is an unavoidable and essential part of design. With this article I wanted to show that studying classical political theories can provide insights about why design is a political activity. I hope to encourage design professionals, researchers, and students to explore this topic in more depth by themselves.

References

1. Löwgren, J. and Stolterman, E. Thoughtful Interaction Design: A Design Perspective on Information Technology. The MIT Press, 2004.
2. Roberts, L. and Baldwin, J. Visual Communication: From Theory to Practice. Fairchild Books AVA, 2006.
3. Alexander, J. Notes towards a definition of politics. Philosophy 89, 348 (2014), 273–300.
4. Collingwood, R.G. Essays in the Political Philosophy of R.G. Collingwood. D. Boucher, ed. Clarendon Press, Oxford, 1989.
5. Collingwood, R.G. The New Leviathan. Oxford Univ. Press, 1942.
6. Oakeshott, M. On Human Conduct. Oxford Univ. Press, 1975.
7. Arendt, H. The Human Condition (2nd Edition). Chicago Univ. Press, 1998.
8. Arendt, H. The Promise of Politics (Reprint Edition). Schocken, 2007.
9. Schmitt, C. The Concept of the Political. Univ. of Chicago Press, 1996.
10. Rancière, J. Dissensus: On Politics and Aesthetics. A&C Black, 2010.
11. Sanders, L. and Stappers, P.J. From designing to co-designing to collective dreaming: Three slices in time. Interactions 21, 6 (2014), 24–33.
12. Brooks, F.P. The Design of Design. Addison-Wesley Professional, 2010.
13. Lawson, B. How Designers Think (Fourth Edition). Architectural Press, 2005.

III NEW IDEAS

10. Experiential Learning of Computing Concepts

Doing research in practice is based on learning from experience. Consequently, the experiential-learning domain can provide us with useful insights about how to effectively learn from experience. In this essay, I explore the experiential learning theory, and present some of lessons learned in applying experiential-learning paradigm in education.[1]

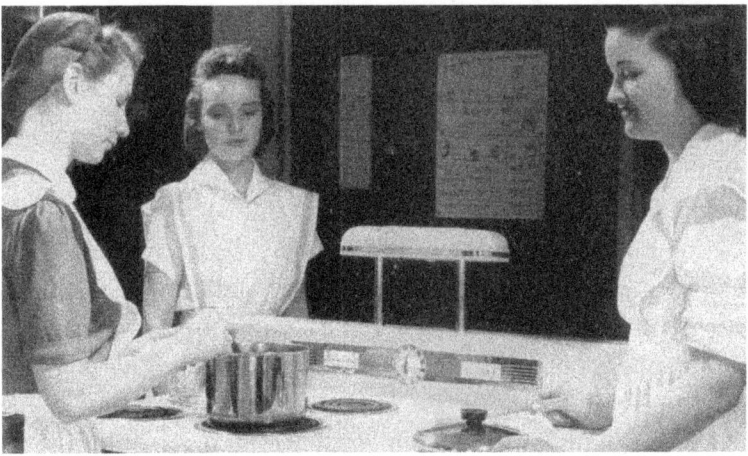

Shimer College students learning to cook by cooking, 1942.

---

[1]This chapter is based on the article Rethinking HCI education: teaching interactive computing concepts based on the experiential learning paradigm, interactions 19, 3 (May 2012), 66-70.

Introduction

Interactive computing technologies such as sensors, actuators, and interactive graphical displays have become increasingly common in cars, household equipment, and other consumer products. As such, industrial and product design professionals, traditionally concerned with the physical form and material properties of products, must now take into account issues related to these interactive technologies. These designers need to understand the possibilities and limitations of computing technologies at a sufficient level to be able to engage in a constructive discussion with computing professionals and to be able to create feasible concept proposals for products that use this technology.

Many design schools have begun to introduce courses on computation to prepare students for these new challenges. These approaches are usually based on adapting and simplifying courses developed in computer science schools, such as teaching students the basics of programming, or introducing the general principles of a particular computing technology. With current tools, however, students generally cannot develop a sufficiently good understanding of the capabilities of computing technologies unless they, themselves, are very skilled programmers or developers. In practice many students do not succeed in mastering the syntax of programming. Computing concepts are often introduced with activities (such as generating lists of prime numbers and making simple line drawings) that are not connected to students' interests or experiences. Additionally, such approaches do not recognize that two radically different education models need to be bridged. Design and craft schools generally follow the experiential learning paradigm, in which knowledge is acquired mainly through doing and working on practical projects [1]. Computer science education, on the other hand, has its roots in mathematics, often emphasizing formal methods and models, articulation of general principles, and a top-down approach to problem solving.

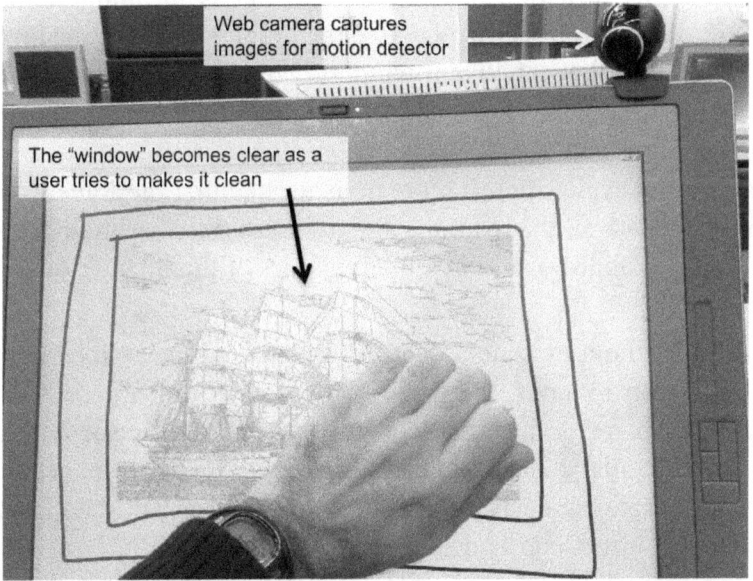

Figure 1. In this example a transparency of a window changes in response to the estimated intensity of hand motion. A motion detector is used to control the transparency of the image representing the window.

Here we discuss our experiences applying a new educational framework for teaching advanced computing concepts compatible with the practice-oriented educational models used in design and craft schools. Figure 1 provides an illustration of our approach. This example shows an exploration of an "intelligent window" concept, where the visibility of the window is changed by "cleaning" it with a hand gesture. Through this example, we want to illustrate several innovations we have started to introduce in the education of interaction designers:

- The student explores a simplified version of a complex computing component, in this case a camera-based motion detector, directly experiencing its possibilities and limitations without needing to know its technical details.
- The exploration and understanding of technology happens in action, in the context that is meaningful for students, directly related to the problem they are dealing with (the "intelligent window" concept), and connected with familiar tools, such as sketching tools and graphical editors.
- The exploration is more holistic, enabling the student to reflect on the relations among user issues, technological possibilities, and overall dynamics of the interaction.

Even without providing detailed understanding, such experiences can pinpoint the limitations of a technology, such as the need for a clear visual field between the sensor and a user's hands, the influence of lighting on the performance of the sensor, the delay caused by the processing of data, and some indirect consequences, such as the user fatigue when the interaction is prolonged. In early phases, this can lead the student to create solutions that overcome these limitations, such as clever positioning of the sensors, adding lighting elements, and making the interaction sessions short enough to avoid user fatigue. In our educational framework, this direct experience of exploring computing technologies is a starting point of the learning process, enabling students to come up with an understanding of computation by reflecting on their experiences.

Background

Experiential learning is a guided process of questioning, investigating, reflecting, and conceptualizing based on direct experiences. In this learning process, the learner is actively engaged, has freedom to choose, and directly experiences the consequences of their actions. There are several models of the experiential learning process, including Kolb's cyclical learning process [2], Schön's reflective practice model [1], Joplin's action-reflection cycle [3], Kesselheim's learning process [4], and Dewey's three-stage process of learning [5]. Though there are differences among these models, the nature of experiential learning is fairly well understood and agreed upon, and all experiential learning models share the following elements [6]:

- actions that create an experience,
- reflections on the action and experience,
- abstractions drawn from the reflections, and
- application of abstractions to a new experience.

Particularly relevant for our work is the experiential learning model developed by Donald Schön. It is one of the most influential and widely accepted models in design- and practice-oriented schools [1]. This model, sometimes also called "reflective practice," stresses the dynamic, cyclic, and reflective nature of design, in which practitioners approach the solution in cycles. In each cycle they interactively frame the problem, generate moves toward a solution, and reflect on the outcomes of these moves.

The Framework

We have begun to develop a framework for teaching advanced computing concepts based on the experiential learning paradigm. With our framework we wanted to enable industrial design students to experience the design of systems that employ advanced computing technologies, such as speech- and camera-based sensors, or Web services, and to learn from that experience. More specifically, we had the following goals:

- To empower students to explore computing technologies without intensive programming. Most of our students are not programmers, and creating systems that employ advanced computing technology using conventional programming languages is beyond their reach.
- To increase students' awareness of the possibilities, limitations, and complexity of computing systems. Many of our students are not aware of the availability and opportunities of emerging computing technologies, and they often have unrealistic expectations about technologies and their complexity.

Having in mind these goals and the discussion about the previous work, we adopted several guiding principles for development of our framework: * We follow the general philosophy behind experiential-based learning: When what we experience differs from the expected or intended, disequilibrium results and our adaptive (learning) process is triggered. Reflection on successful adaptive operations (reflective abstracting) leads to new or modified concepts. The challenge is to create learning environments that are complex enough to lead to unexpected experiences, but not too complex to be inaccessible to students. * Unguided or minimally guided experience and reflection is not effective [7]. We must provide a structure and a set of plans that support the development of informed exploration and reflective inquiry without taking initiative or control away from students. We need to create a personally meaningful context for students. For a problem to foster the learning of powerful computing ideas, the students must accept it as their problem. We need to take noncontextualized computing ideas and embed them in a meaningful context for student investigation.

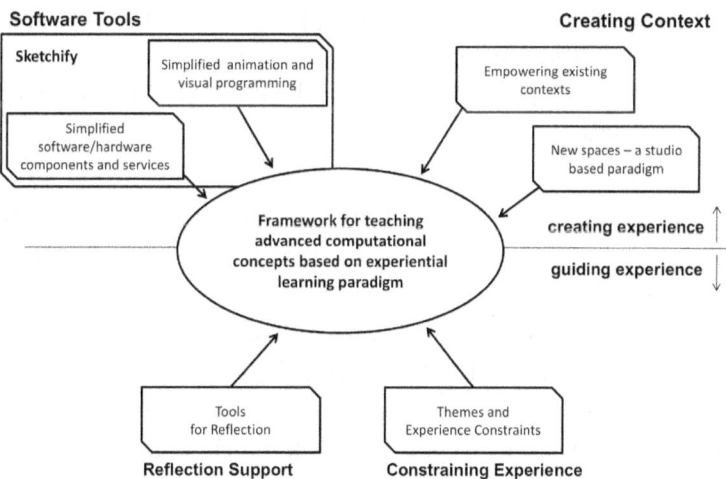

Figure 2. The framework for experiential learning of computing concepts, consisting of tools that facilitate creation of useful experiences and tools that guide and support reflection on such experiences.

Our framework supports these principles with a collection of software tools, conceptual frameworks, and guidelines, which we classify into two groups (Figure 2):

- tools that facilitate the creation of useful experiences for exploring advanced computing technology, and
- tools that guide and support reflection on such experiences.

Creating experience. The key element of our approach is empowering students to have relevant experiences with advanced computational technologies, because without such experiences, the students do not have a basis to reflect and learn. To support this goal, our framework includes:

- a collection of software tools to empower students to explore the limitations and opportunities of technologies without intensive programming, and
- tools and spaces for creating the contexts for such experiences.

Existing approaches to teaching computing concepts do not usually allow exploration of advanced computing technologies unless a student is willing to become a skilled programmer and learn a significant amount of technological detail. Our goal was to facilitate creating learning experiences, as described in our introductory example, without requiring students to program or to obtain detailed technical knowledge.

The main software tool for our educational activities was the Sketchify toolkit (http://sketchify.sf.net). This software toolkit helps non-programmers build systems with complex computing technologies; it served as the basis for supporting the experiential learning in our courses. In this way, we could bring components from various domains within students' reach, allowing them to directly experience possibilities and limitations of technologies without needing advanced programming skills. Sketchify enables students to combine these technology samples with drawing tools and simple end-user programming techniques, such as spreadsheets. We have incorporated many samples of computing technologies within Sketchify, including text-to-speech engines and speech recognizers, Web services (such as the Google search engine), Phidgets, Arduino, semantic services (such as the Wordnet definition service), camera-based face and motion detectors, MP3 and MIDI players, Wii Remote, a Car Simulator, and many others. (A detailed description of the tool is available in [8]).

In addition to using and developing software tools that can enable students of diverse backgrounds to explore advanced computing technology, we have been working on creating contexts for such experiences. Sketchify focuses on enabling students to exploit the computing technologies available in their environment by allowing them to use everyday computing objects and artifacts, such as their mobile phones, game devices, cameras, or microphones. We developed a number of Sketchify adapters for these objects.

We are also working on creating spaces with specialized equipment and a more stimulating atmosphere. One such space is the ConceptLab, a studio that reflects our vision of what a design studio of the future could look like.

Guiding experience. With the tools described earlier, we can empower students to engage in useful learning experiences. In reference to our second general principle, which states that unguided experience is not effective, we also developed several conceptual tools that can help educators guide and structure students' experience and reflection.

In all of our educational activities, we asked students to keep a creative logbook in which to write down what they had learned and to reflect on the techniques they were using. We also encouraged public discussions, not as mere presentation activities, but as an opportunity to reflect on the experience and learn something new.

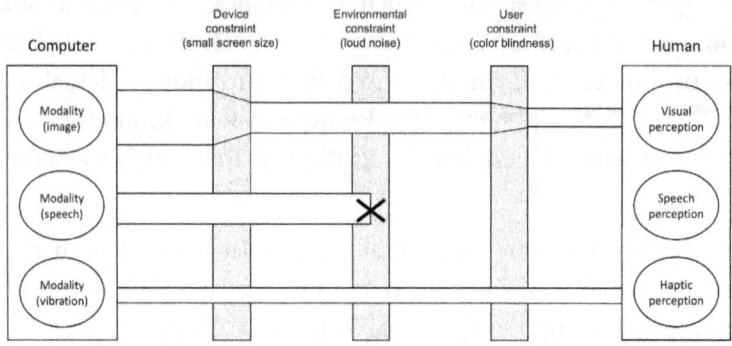

Figure 3. An example of reflection about interaction through interaction constraints. Presenting information through image modality is limited by screen size and users' color disability. Speech may not be perceived in very noisy environments. Using vibrations, such as on a mobile phone, is not affected by the device screen size, environmental noise, or user color blindness, although it has limited information bandwidth.

To help guide students reflect on discussions, presentations, and notes, we provided several structured frameworks. The main purpose of these frameworks is to give students a structure to reflect on their experiences and to provide them with a shared vocabulary they can use to critically review each other's work. A systematic analysis and reflection on concrete systems can reveal potential problems and inspire new features. Frameworks for reflection also provide a way to introduce and give meaning to computing concepts. For example, in courses related to the design of interactive systems, we used an adaptation of the framework for modeling human-computer interaction in terms of interaction constraints [9]. This model is presented in Figure 3. The idea behind this modeling framework is that an interactive system could be described in terms of requirements that it imposes on users, such as usage of the visual field or audio perception. This description is discussed in terms of potential constraints that may influence the interaction, such as device limitations, user (dis)abilities, and the environmental influence.

Another important component of our framework is experience constraints. While we stimulate students to work on their own problems and set their own goals, when working in groups we found it useful to give a direction to students' activities. Rather than setting a concrete goal, experience constraints are aimed at giving the "mood" to the whole educational setting and activities. Experience constraints thus serve two roles (Figure 4):

- constrain student explorations, providing inspiration, giving direction, and focusing students' activities; and
- provide a unified theme for student actions and projects, in order to facilitate communication among students.

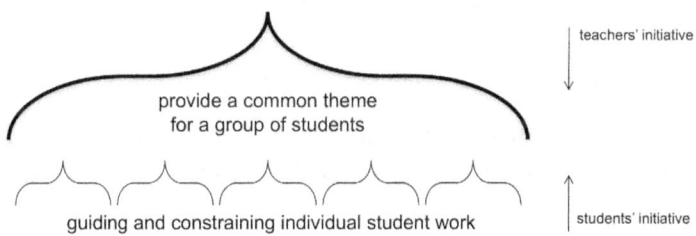

Figure 4. The role of experience constraints in our framework.

Existing approaches to teaching computing concepts do not usually allow exploration of advanced computing technologies unless a student is willing to become a skilled programmer and learn a significant amount of technological detail.

We used various themes and metaphors to provide experience constraints. For example, in our master's course on multimodal interaction we used the "Power Trio" theme to inspire and unify students' activities. In our undergraduate course "Sketching Interactive Systems" we used the theme of sketching to encourage them to explore more diverse technologies.

Our experience constraints have a role similar to that of a primary generator in design used to *"narrow down the space of possible solutions by providing an initial focus, i.e., by constraining and guiding the designer's development of a solution"* [10].

Conclusion

Our framework has been developed and applied during a period of three years at the Department of Industrial Design at the Eindhoven University of Technology. We used it in three iterations of the undergraduate course "Sketching Interactive Systems" and three iterations of the postgraduate course "Multimodal Interaction." The first results are encouraging, and although it's still too early to make more specific claims, our initial findings suggest the following:

The key element of our approach was to empower students to have relevant experiences with advanced computational technologies. Without such experience, the students do not have a basis to reflect and learn. This was a particularly successful element in the usage of our framework, especially in undergraduate education.

Our tools enabled students to discover and learn a range of important properties of current computing technologies, as well as some basic computing abstractions, such as variables. Though such experience has its limits and cannot enable students to discover all relevant concepts, it provides a productive context to discuss such concepts and increases the general interest of students.

Providing a structure and a set of plans that support informed exploration and reflective inquiry was crucial to enabling students to learn from their experience and from each other. Simply letting students explore computing technology and build computational systems will not necessarily help them learn computing concepts.

Having themes and constraining students' experiences had a positive effect on the conceptual integrity of our educational activities and on student collaboration. However, the theme has to be introduced carefully and clearly to avoid confusion among students about its role.

References

1. Schön, D.A. The Reflective Practitioner. Basic Books, New York, 1983.
2. Kolb, D.A. Experiential Learning: Experience as the Source of Learning and Development. Prentice Hall, New Jersey, 1984.
3. Joplin, L. On defining experiential education. Journal of Experiential Education 4, 1 (1981), 17–20.
4. Kesselheim, A.D. A rationale for outdoor activity as experiential education: The reason for freezing. Proc. 1st North American Conference on Outdoor Pursuits in Higher Education (Boone, NC). 1974, 18–22.
5. Dewey, J. Experience and Education. Simon and Schuster, New York, 1938/1997.
6. Stehno, J.J. The application and integration of experiential education in higher education. Touch of Nature Environmental Center, Southern Illinois University, Carbondale, IL, 1986; (Eric Doc. Reproduction Service No ED-285-465).
7. Kirschner, P.A., Sweller, J., and Clark, R.E. Why minimal guidance during instruction does not work: An analysis of the failure of constructivist, discovery, problem-based, experiential, and inquiry-based teaching. Educational Psychologist 41, 2 (2006), 75–86.
8. Obrenovic, Ž. and Martens, J.B. Sketching interactive systems with Sketchify, ACM Transactions on Computer Human Interaction 18, 1 (March 2011), Article 4.
9. Obrenovic, Ž., Starcevic, D. and Abascal, J. Universal accessibility as a multimodal design issue. Commun. ACM 50, 5 (May 2007), 83–88.
10. Lawson, B. How Designers Think: The Design Process Demystified (4th ed.). Architectural Press, 2005.

11. Sketchifying: Bringing Innovation into Software Development

Doing research in practice also requires intensive experimentation. Software Sketchifying is an idea about how to bring a research mindset and more experimentation in software development.[1]

Mr. and Mrs. Henry Ford in his first car, the Ford Quadricycle.

---

[1]This chapter is based on the article Software Sketchifying: Bringing Innovation into Software Development, IEEE Software 30, 3 (May-June 2013), 80-86.

Introduction

Henry Ford's assembly-line production of the Model T inspired changes in the automotive industry, and the software industry has made numerous attempts to apply similar ideas (for example, see the chapter, "Will the Real Henry Ford of Software Please Stand Up" in Robert L. Glass's book) [1]. While the assembly-line philosophy is well known, Ford's approach to innovation and the process that preceded the Model T's production is less so. Between 1892 and the formation of the Ford Motor Company in 1903, while working mostly for the Edison Illuminating Company, Ford built about 25 cars. In the five years after the company's formation, he built and sold eight models—Models A, B, C, F, K, N, R, and S— before settling on the Model T. He tested prototypes labeled with the 11 missing letters. Ford summed up this experience this way [2]:

I do not believe in starting to make until I have discovered the best possible thing. This, of course, does not mean that a product should never be changed, but I think that it will be found more economical in the end not even to try to produce an article until you have fully satisfied yourself that utility, design, and material are the best. If your researches do not give you that confidence, then keep right on searching until you find confidence.... I spent twelve years before I had a Model T that suited me. We did not attempt to go into real production until we had a real product.

Today's automotive industry has changed significantly since Ford's initial success, but some of his philosophy behind innovation still remains. For example, Toyota's "nemawashi" principle states that decisions should be implemented rapidly but made slowly, by consensus, and after considering all options[3]. Bill Buxton, who studied innovation in the automotive industry, noted that a new car's design phase starts with a broad exploration that culminates in the construction of a full-size clay model and costs over a quarter of a million dollars[4]. Only after bringing the new concept to a high level of fidelity in terms of its form, business plan, and engineering plan does a project get a "green light." After that, it typically takes a year of engineering before the project can go into production.

Inspired by general ideas about how the automotive industry brings innovation into manufacturing, I developed software sketchifying as an activity to stimulate and support software stakeholders to spend more time generating and considering alternative ideas before making a decision to proceed with engineering. My view on software sketchifying combines general ideas of sketching[4] and creativity support tools[5] with several existing software engineering approaches. To support and explore this view, I developed Sketchlet (http://sketchlet.sourceforge.net), a flexible, Java-based tool that empowers engineers and nonengineers to work with emerging software and hardware technologies, explore their possibilities, and create working examples—called sketchlets—that incorporate these emerging technologies.

Product Innovation and Software Engineering

Contrary to the automotive industry, the software industry has a rich history of engineering wrong products. Ill-defined system requirements and poor communication with users remain top factors that influence software project failures[6]. Frederick Brooks also noted that the hardest single part of building a software system is deciding precisely what to build[7]. He proposed rapid system prototyping and iterative requirements specification as a way to solve this problem. Many existing software engineering methodologies, including the Rational Unified Process, Extreme Programming, and other agile software development frameworks follow iterative and incremental approaches.

However, these approaches have limitations when it comes to true innovation. Although prototyping can let us cheaply represent and test our ideas, and iterative and incremental development can help further refine our ideas based on frequent user feedback, neither approach directly supports the generation of new product ideas, nor do they encourage the consideration of alternatives.

Buxton went further in his critique of the innovation capacity of iterative, incremental software development, seeing no comparison between software product design and the development of new automobiles[4]. He argued that innovative software projects need at least a distinct design phase followed by a clear green-light process before proceeding to product engineering. He saw design and engineering as different activities that employ different processes and for which people suited to one are typically not suited for the other.

A Sketchifying Scenario

Consider an example scenario with Mirko, an interaction designer at a company that builds software for new generations of cars with advanced sensing and display technologies. Mirko has recently joined the company to explore ideas for software applications that exploit novel opportunities, such as using data from a car radar, GPS sensors, and links to Web services.

Mirko's first task is to explore two applications: a system for warning about the proximity of other cars and a system for presenting news in idle situations, such as waiting for a traffic light. Mirko isn't a programmer, nor is he familiar with all the technical possibilities of modern cars, but he uses a design environment through which he can access and explore software services and components related to his task without serious programming.

To understand what's possible, Mirko first talks with several of his company's engineers. They advise him to start by using a car simulator, which provides a realistic but safe environment to learn about new automotive technologies. One engineer also writes a small adapter that connects the car simulator logger to Mirko's design tool. This adaptation gives Mirko immediate access through a simple spreadsheet-like interface to the simulator data—such as the car's speed and its distance from the car in front of it. Mirko starts a design environment on his laptop and connects it to the simulator. After becoming familiar with the simulator's possibilities, he turns to his laptop to create a few sketchlets, which are simple interactive pieces of software.

Proximity Warning System. To explore the options for implementing a proximity warning system, Mirko first considers three presentation modes: graphical, audio, and haptic (vibration). For graphical presentation, he uses an editor in his design environment and creates several simple drawings. Then he opens the properties panel and connects the variables from the car simulator to the graphical properties of drawn regions. For example, he creates a sketchlet in which an image's transparency dynamically changes as a function of the distance from the car in front of the driver. He then experiments with other graphical properties, such as image size, position, or orientation. He returns to the simulator and tries each alternative. He also invites a few colleagues to try out and comment on his ideas.

After exploring graphical options, he proceeds to create audio sketchlets. He first tries a MIDI-generator service and connects data coming from the sensor to MIDI note parameters, such as pitch or tone duration. He also experiments with a text-to-speech service, generating speech based on the conditions derived from car data. Finally, he explores using an MP3 player with prerecorded sounds. He then goes back to the simulator and tries these alternatives.

Mirko also wants to try a vibration modality to present navigation information, which the simulator doesn't support. He decides to use a simple trick, starting an application on his mobile phone that lets his design environment control the phone's resources, including its vibrator. Using gaffer tape, he fixes the mobile phone to the steering wheel and creates several sketchlets that map the distance from the car in front of him to vibration patterns. Marko knows it's not a very elegant solution, but it lets him explore basic opportunities of this modality with available resources and little work.

News Presentation. Mirko also plays with some other options related to the application for presenting news. He starts a Google news service in his design environment and creates a simple page that presents an HTML output of the news service. He then creates a condition for the page's visibility so that the news appears as an overlay on part of the windshield, but only when the car's speed is zero and the automobile is not in gear. He also experiments with speech services that let a user set a news search query by speech.

After finishing his work in the lab, Mirko decides to collect some real-world experiences and try some of his more promising sketchlets in a real car. With help from engineers who are working on testing cars, Mirko gets an extension of his design environment that uses a Bluetooth connection to a test car's on-board diagnostic (OBD) system. With this addition, Mirko creates a simple setting using his smartphone as a presentation device, positioned under a windshield. He connects the smartphone to his laptop, which uses a simple remote desktop client to capture a part of a screen from his laptop. On the laptop, Mirko is running the sketchlets that he created in the lab and that are now connected to the car's OBD system. He asks a colleague to drive the car while he observes a situation and videorecords a whole session for later analysis.

During the process, Mirko constantly interacts with other stakeholders, regularly presents his findings, and lets clients and colleagues try out some of his sketchlets. In this way, Mirko is helping develop new products by providing realistic and tested ideas before and outside the main development activity.

Software Sketchifying

I built on Buxton's suggestion by introducing software sketchifying into software product development as a complement to prototyping and engineering. The sidebar presents a sketchifying example scenario of how it might work in developing software systems for an automobile.

Software Sketchifying Approach. One key characteristic of this approach is postponing the main development activity for the benefit of free exploration, following a main principle of creativity: to generate a good idea, you must generate multiple ideas and then dispose of the bad ones. 1,4 Another key characteristic is stimulating early involvement of nonengineers. Such users often have expertise that's important for understanding customers and their needs. More specifically, the example scenario in the sidebar illustrates several points about software sketchifying:

- The designer's main activity is exploration, learning about a problem and potential solutions and answering a question about what to build.
- Such explorative activity is heuristic, creative, and based on trial and error, rather than incremental and iterative. The designer generates several ideas, most of which will be rejected. However, this process yields important lessons and stimulates generation of novel ideas. These lessons and ideas are the activity's main outcome.
- The exploration activity is not accidental, but disciplined and systematic.
- The exploration is holistic, enabling designers to reflect on relations among user issues, software and hardware possibilities, and the overall dynamics of human-computer interaction. The ideas in the example scenario are influenced not only by software but also by human factors and problems related to car mechanics and equipment.
- The exploration enables early user involvement through

simple but functional pieces of software in the form of sketchlets.
- Working with real systems, such as the car simulator, car diagnostic system, and Web services, lets a designer learn about the possibilities and limitations of software technologies and create conceptual proposals that are more realistic.

Designers generally aren't engineers who can program and extend their design environments. However, they're part of a broader community of people who can help them learn and extend the exploration space on an ad hoc basis. Sketchifying supports this interaction without taking too much time, thereby empowering nonengineers to explore emerging technologies and to test their ideas without additional help from developers.

Software Sketchifying Tools. To support and explore this approach, Sketchlet combines elements from traditional sketching, software hacking, opportunistic software development, and end-user development. Sketchlet builds on the results of the Sketchify project (http://sketchify.sourceforge.net), which explored possibilities to improve early design stages and education of interaction designers. 8

Sketchlet has two main roles:

- to enable designers to create a number of simple pieces of software—sketchlets—as a way to quickly and cheaply explore software and hardware technologies and their potential applications, and
- to support involvement of software engineers in short, ad hoc sessions that give designers realistic pieces of technologies that might be useful for design exploration.

Sketchlet lets designers interact directly with software and hardware technologies through a simple, intuitive user interface. To simplify the integration with these technologies, Sketchlet combines ideas from opportunistic software development with techniques used by hacking and mashup communities[9, 10]. A full description of Sketchlet is out of scope for this article (see sketchlet.sf.net[2] for details).

Initial Sketchlet Applications and Results

I've developed and applied the ideas about software sketchifying in three projects that featured collaboration among software engineers, interaction designers, and researchers. In these projects, interaction designers and researchers were primarily responsible for creating and evaluating novel conceptual proposals and ideas:

- *Connect & Drive* project[3]. Several researchers used Sketchlet to explore options for developing software systems for cooperative adaptive cruise control systems in cars, based on Wi-Fi communication between vehicles and road infrastructure.
- *Persuasive Technology, Allocation of Control, and Social Values* project. Sketchlet played a similar role as it did in the Connect & Drive project, helping researchers investigate software products for developing persuasive technologies that encourage people to hand over control to intelligent automation of cars.

---

[2]http://sketchlet.sf.net
[3]https://www.tue.nl/en/university/departments/industrial-design/research/research-groups/user-centered-engineering/research/projects/explorations-in-interactions/connect-drive/

- *RePar* project (Resolving the Paradox of User Centered Design[4]). Sketchlet was one of the flexible prototyping tools in user-centered design processes, allowing designers to create and evaluate (ill-defined) product concepts early in the development.

Although Sketchlet is still in early development, the approach and tool showed several positive effects in these projects. First, it broadened the opportunities to constructively involve nonengineers, including interaction designers, psychologists, and students. Our tools empowered nonengineers to easily explore relevant technologies and to independently create and test their ideas. The companies involved benefited from their nonengineering expertise and knowledge early in the design process.

Sketchlet also promoted different collaboration between engineers and nonengineer designers. Prior to using Sketchlet, most of the companies followed the approach of making designers responsible for creating a conceptual proposal, which they gave to developers for implementation with little interaction, except to clarify their designs. With Sketchlet, the interaction between designers and engineers could work in two ways, with engineers giving designers simplified versions of software components and services—early in the design process—that the engineers might use later in the implementation (see Figure 1).

---

[4]https://www.utwente.nl/ctw/opm/research/hcd/projects/Resolving%20the%20paradox%20of%20user-centred%20design/

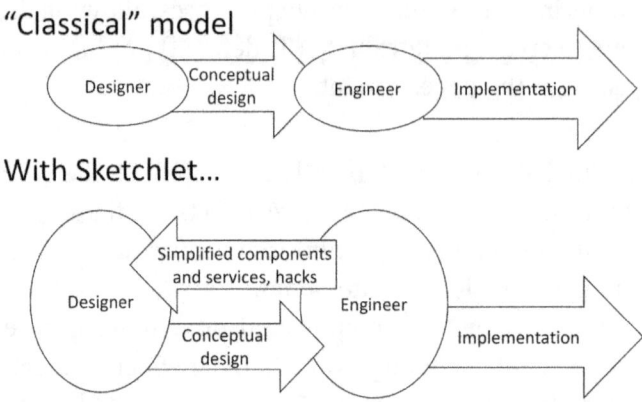

Figure 1. Comparing the classical design-engineering interaction with sketchifying. With sketchifying, supported by tools like Sketchlet, the interaction between designers and engineers can work in two ways, allowing engineers to give designers early access to simplified versions of software components and services that the engineers might use later in the implementation.

The connected services, although simplified, resemble real components, and sketchlets expressed in terms of these services come closer to the implementation platform that the engineers will use. This change addressed one problem that many companies experience when designers and engineers need to work together—namely, the engineers perceive designers' ideas as unrealistic, too distant from available technology, and not precise enough to be useful. Through the exploration of these services, designers can develop more realistic expectations about the possibilities and limitations of technologies, and incorporate this understanding into design proposals.

Lastly, Sketchlet influenced the mindset of companies toward more and broader explorations early in the software design. Sketchlet helped illustrate the potential of such exploration and inspire the companies to think how other tools could be used in a similar explorative way.

Sketchifying Benefits and Relation to Other Approaches

Software sketchifying can help better define product requirements so that the subsequent engineering process has a clear focus and goal. It promotes direct exploration of emerging technologies and creation of working examples of simple pieces of software with these technologies as a way to identify potential problems and provoke reactions from users as early as possible. The tool shows the effects of design decisions on user experience and supports user testing before actual development starts.

Exploring the possibilities and limitations of technologies early in the design helps identify a number of problems or user issues before investing in a significant development effort. Discovering such problems later in the process could require changes and additional effort. Early discovery is particularly important in projects using emerging technologies, which have many unknowns—including how well users will accept them.

Promoting the constructive involvement of nonengineers in the design process opens the door to help from experts in fields such as human psychology, which in turn reduces the burden on developers. Moreover, as Glass noted [1], users who understand the application problem to be solved are often more likely to produce innovation than computer technologists, who understand only the computing problem to be solved. The sketchifying approach requires occasional involvement of developers, but it aims to incorporate them in short ad hoc sessions, and the intent is to empower nonengineers to explore further without developers' help. Once the developer adapts some technology for Sketchlet, nonengineers can work with this technology through a simple end-user interface that does not require technical expertise or programming knowledge.

Relation to Prototyping and Engineering

Software sketchifying complements existing prototyping and engineering approaches by its focus on free exploration and a trial-and-error approach versus a more iterative, incremental approach of prototyping and engineering (see Figure 2).

Sketchifying supports users in constructing a novel idea and enables nonengineers to actively contribute. This brings software design closer to the practice of other engineering disciplines, in which the design phase precedes the main engineering activity, and designers (usually nonengineers) are encouraged to freely explore ideas before consolidating a few of them for further development. For instance, it's not unusual for an industrial designer to generate 30 or more sketches a day in the early stages of design, each possibly exploring a different concept [4].

Software sketchifying precedes prototyping, which tests, compares, and further develops aspects of selected ideas. With a prototype in place, the development can take an evolutionary approach. Prototyping should assess whether selected ideas are feasible and should help decide whether to proceed with engineering. Prototyping aims at making an idea more detailed and concrete, rather than coming up with radically new ideas. Engineering turns the winning idea into a robust and usable product.

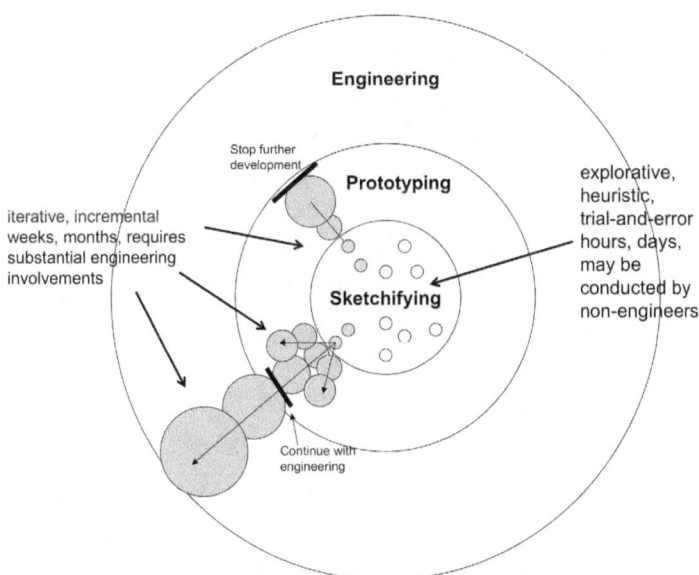

Figure 2. An idealized representation of relationships among sketchifying, prototyping, and engineering. Sketchifying supports users in constructing a novel idea. It precedes prototyping, which tests, compares, and further develops aspects of selected ideas. Engineering turns the winning idea into a robust and usable product.

Relation to Other Software Tools

In principle, tools other than Sketchlet could implement the sketchifying idea. However, many current tools can't fully support it because they're not optimized for free exploration and involvement of nonengineers. For example, we could use standard programming languages, such as Java, C#, C++, or programming tools oriented toward interaction design such as Flash and Processing to implement our example scenario. However, programming requires significant expertise, time, and effort—an investment that's simply too high for the intended purpose of generating new ideas and exploring possibilities.

Existing low-fidelity prototyping environments provide ways to quickly create prototypes with inputs taken from external services or sensors. 11,12 These environments might be excellent choices for exploring interactions in various domains. The problems I'm addressing cross these domains and require a variety of sensory inputs and links to diverse software services as well as additional components specific to the companies I'm working with. In addition, such tools often require too much precise specification, partly because they're primarily developed for advanced prototyping rather than for free and broad exploration.

Electronic sketching systems are another promising direction for design tools, enabling designers to create interactive systems with ease using intuitive and natural pen gestures [13]. From the viewpoint of my example scenario, these systems have the drawback of being specialized for specific domains and used successfully only in inherently graphical domains that have a stable and well-known set of primitives, such as 2D and 3D graphics or websites.

Another alternative is to use simple freehand drawings and techniques such as screen prototyping. Such techniques can help in exploring a solution's graphical elements. However, they can describe overall system interactions, such as sensing device inputs and user response dynamics, only in very abstract terms. Moreover, paper sketching doesn't let users explore the possibilities and limitations of emerging technologies. Direct exploration of such technologies yields more concrete ideas about how to best employ them.

Sketchlet borrows ideas from existing solutions, while trying to overcome some of their limitations. I also see it as a complement to existing tools, rather than a replacement. On several occasions, designers have used Sketchlet in conjunction with other tools. For example, some of our users employed Max MSP for signal processing and audio effects and Sketchlet for connections to sensor devices and visualization.

My initial experiences with applying software sketchifying are encouraging. However, an important limitation of this approach is that it requires significant changes of current development culture in its emphasis on postponing the start of development to benefit free exploration, more active involvement of nonengineers and end users, and new forms of interaction between engineers and nonengineers prior to the main development activity. Such changes, in my experience, aren't easy to achieve, but without them, the sketchifying tools are less effective and tend to be used in a limited way.

Sketchifying: Bringing Innovation into Software Development

Henry Ford with Model T, 1921

References

1. R.L. Glass, Software Creativity 2.0, developer.* Books, 2006.
2. J. Grudin, "Travel Back in Time: Design Methods of Two Billionaire Industrialists," ACM Interactions, vol. 15, no. 3, 2008, pp. 30–33.
3. J. Liker, The Toyota Way: 14 Management Principles from the World's Greatest Manufacturer, McGraw-Hill, 2004.
4. B. Buxton, Sketching User Experiences: Getting the Design Right and the Right Design, Morgan Kaufmann, 2007.
5. B. Shneiderman, "Creativity Support Tools: Accelerating Discovery and Innovation," Comm. ACM, vol. 50, no. 12, 2007, pp. 20–32.
6. R.N. Charette, "Why Software Fails," IEEE Spectrum, vol. 42, no. 9, 2005, pp. 42–49.
7. F. Brooks, "No Silver Bullet—Essence and Accidents of Software Engineering," Computer, vol. 20, no. 4, 1987, pp. 10–19.
8. Ž. Obrenović and J.B. Martens, "Sketching Interactive Systems with Sketchify," ACM Trans. Computer-Human Interaction, vol. 18, no. 1, 2011, article 4.
9. B. Hartmann, S. Doorley, and S.R. Klemmer, "Hacking, Mashing, Gluing: Understanding Opportunistic Design," IEEE Pervasive Computing, vol. 7, no. 3, 2009, pp. 46–54.
10. Ž. Obrenović,D. Gaševic, and A. Eliëns, "Stimulating Creativity through Opportunistic Software Development," IEEE Software, vol. 25, no. 6, 2008, pp. 64–70.
11. M. Rettig, "Prototyping for Tiny Fingers," Comm. ACM, vol. 37, no. 4, 1994, pp. 21–27.

12. Y.K. Lim, E. Stolterman, and J. Tenenberg, "The Anatomy of Prototypes: Prototypes as Filters, Prototypes as Manifestations of Design Ideas," ACM Trans. Computer-Human Interaction, vol. 15, no. 2, 2008, article 7.
13. J.A. Landay and B.A. Myers, "Sketching Interfaces: Toward More Human Interface Design," Computer, vol. 34, no. 3, 2001, pp. 56–64.

To Probe Further: Selected Bibliography

Design Community

Schön D. (1983):
The Reflective Practitioner,
London: Temple Smith.

Simon H.A. (1996):
The Sciences of the Artificial,
The MIT Press; 3rd edition.

Lawson B. (2005):
How designers think,
Architectural Press, 4th edition.

Alexander C. (1964):
Notes on the Synthesis of Form,
Harvard University Press.

Cross N. (2006):
Designerly Ways of Knowing,
Springer.

Frayling C. (1993):
Research in Art and Design,
Royal College of Art Research Papers 1, pp. 1-5.

Hales, C. (1987):
An Analysis of the Engineering Design Process in an Industrial Context, Gants Hill, 1987.

Rust C., Mottram J., Till J. (2007):
AHRC Research Review: Practice-Led Research in Art, Design and Architecture[5],
Elshaw.

Horvath, I. (2001):
A contemporary survey of scientific research into engineering design[6],
13th International conference on engineering design, pp. 13-20.

Horváth, I. (2007):
Comparison of three methodological approaches of design research[7],
International Conference on Engineering Design, ICED'07.

Stolterman, E. (2008):
The nature of design practice and implications for interaction design research[8],
International Journal of Design 2(1), pp. 55-65.

Stappers, P. J. (2007):
Doing Design as a Part of Doing Research[9],
In: Michel R, ed. *Design Research Now: Essays and Selected Projects: Birkenhauser*, pp. 81–91.

McNiff S. (2008):
Art-Based Research[10],
in J. Gary Knowles & Ardra L. Cole (Eds.) *Handbook of*

---

[5]http://arts.brighton.ac.uk/__data/assets/pdf_file/0018/43065/Practice-Led_Review_Nov07.pdf

[6]http://citeseerx.ist.psu.edu/viewdoc/download?doi=10.1.1.142.5463&rep=rep1&type=pdf

[7]http://www.designsociety.org/download-publication/25512/comparison_of_three_methodological_approaches_of_design_research

[8]http://www.ijdesign.org/ojs/index.php/IJDesign/article/view/240/148/

[9]http://dx.doi.org/10.1007/978-3-7643-8472-2_6

[10]http://www.moz.ac.at/files/pdf/fofoe/ff_abr.pdf

the Arts in Qualitative Research: Perspectives, Methodologies, Examples, and Issues

Polanyi, M. (1974):
Personal Knowledge,
Chicago: U Chicago Press.

Polanyi, M. (1983):
The Tacit Dimension,
Peter Smith Publisher.

Practice Research[11], Wikipedia Page

Yee J.S.R. (2010):
Methodological Innovation in Practice-Based Design Doctorates[12],
Journal of Research Practice 6 (2)

Mellor D.H. (2015):
Artists and Engineers[13],
Philosophy 90(3), pp. 393-402

HCI / Interaction Design

Brooks F.P. (1988):
Grasping reality through illusion–interactive graphics serving science[14],
Proc. CHI '88. ACM, New York, NY, pp. 1-11.

Zimmerman, J., Forlizzi, J., and Evenson, S. (2007):
Research through design as a method for interaction design research in HCI[15],
Proc. CHI '07. ACM, New York, NY, pp. 493-502.

---

[11] http://en.wikipedia.org/wiki/Practice_research
[12] http://jrp.icaap.org/index.php/jrp/article/view/196/193
[13] https://doi.org/10.1017/S0031819115000133
[14] http://dx.doi.org/10.1145/57167.57168
[15] http://dx.doi.org/10.1145/1240624.1240704

Zimmerman J., Stolterman E., and Forlizzi J. (2010):
An analysis and critique of Research through Design: towards a formalization of a research approach[16],
DIS '10. ACM, New York, NY, USA, pp. 310-319.

Greenberg S. and Buxton B. (2008):
Usability evaluation considered harmful (some of the time)[17],
CHI '08, 111-120.

Zimmerman J., Evenson S., Forlizzi J. (2004):
Discovering Knowledge in the Design Case[18],
Proc. Future Ground (2004). Design Research Society.

Brenda L. (2003):
Design Research: Methods and Perspectives,
MIT Press.

Koskinen I., Zimmerman J., Binder T., Redstrom J., Wensveen S. (2011):
Design Research Through Practice: From the Lab, Field, and Showroom,
Morgan Kaufmann Publishers.

Wolf T.V., Rode J.A., Sussman J., and Wendy A. Kellogg. (2006):
Dispelling "design" as the black art of CHI[19],
ACM CHI '06, pp. 521-530.

Gaver W. (2012):
What should we expect from research through design?[20],
ACM CHI '12. pp. 937-946.

Basballe D.A. and Halskov K. (2012):

---

[16] http://dx.doi.org/10.1145/1858171.1858228
[17] http://doi.acm.org/10.1145/1357054.1357074
[18] http://www.cs.cmu.edu/~johnz/pubs/2004_FutureGround.pdf
[19] http://dx.doi.org/10.1145/1124772.1124853
[20] http://dx.doi.org/10.1145/2207676.2208538

Dynamics of research through design[21],
DIS '12. ACM, New York, NY, USA, pp. 58-67.

Buxton, W. (2007):
Sketching user experience – Getting the design right and the right design,
San Francisco, CA, Morgan Kaufmann.

Landay J. (2009):
I give up on CHI/UIST[22],
Blog entry, Nov 7, 2009.

Sengers, P. (2006):
*Must design become 'scientific'?[23],
DIS'06 Workshop on Exploring Design as a Research Activity.

Dourish, P. (2006):
Implications for design[24],
Proc. CHI '06. NY: ACM Press, 541-550.

Fallman, D. (2003):
Design-oriented human-computer interaction[25],
Proc. CHI 2003, 225-32.

Gaver, W. (2006):
Learning from Experience: The Humble Role of Theory in Practice-Based Research,
CHI 2006 Workshop on Theory and Methods for Experience-Centered Design.

Goodman, E., Stolterman, E. & Wakkary, R. (2011):
Understanding Interaction Design Practicies[26],

---
[21] http://dx.doi.org/10.1145/2317956.2317967
[22] http://dubfuture.blogspot.nl/2009/11/i-give-up-on-chiuist.html
[23] http://www.antle.iat.sfu.ca/courses/iat834/resources/Sengers_06_design-v-science-death-match.doc
[24] http://dx.doi.org/10.1145/1124772.1124855
[25] http://dx.doi.org/10.1145/642611.642652
[26] http://uxdesignpractice.com/papers/p1061-goodman.pdf

CHI 2011, May 7–12, 2011, Vancouver, BC, Canada

Roedl D.J and Stolterman E. (2013):
Design research at CHI and its applicability to design practice[27],
Proc. CHI '13. pp. 1951-1954.

Rogers, Y. (2004):
New theoretical approaches for human-computer interaction[28],
Annual review of information, science and technology 38 (pp. 87–143).

Blythe M. (2014):
Research through design fiction: narrative in real and imaginary abstracts[29],
Proc. CHI '14. ACM, New York, NY, USA, pp. 703-712.

Dachtera J., Randall D., and Wulf V. (2014):
Research on research: design research at the margins: academia, industry and end-users[30],
Proc. CHI '14. ACM, New York, NY, USA, pp. 713-722.

Dalsgaard P. and Dindler C. (2014):
Between theory and practice: bridging concepts in HCI research[31],
Proc. CHI '14, ACM, New York, NY, USA, pp. 1635-1644.

Claes S., Wouters N., Slegers K., and Moere A.V. (2015):
Controlling In-the-Wild Evaluation Studies of Public Displays[32],

---

[27] http://doi.acm.org/10.1145/2466110.2466257
[28] http://www.informatics.sussex.ac.uk/research/groups/interact/publications/ARIST-Rogers.pdf
[29] http://dx.doi.org/10.1145/2556288.2557098
[30] http://dx.doi.org/10.1145/2556288.2557261
[31] http://dx.doi.org/10.1145/2556288.2557342
[32] http://dx.doi.org/10.1145/2702123.2702353

Proc. CHI '15, ACM, New York, NY, USA, pp. 81-84.

Luusua A., Ylipulli J., Jurmu M., Pihlajaniemi H., Markkanen P., and Ojala T. (2015):
Evaluation Probes[33],
Proc. CHI '15, ACM, New York, NY, USA, pp. 85-94.

Remy C., Gegenbauer S., and Huang E.M. (2015):
Bridging the Theory-Practice Gap: Lessons and Challenges of Applying the Attachment Framework for Sustainable HCI Design[34],
Proc. CHI '15, ACM, New York, NY, USA, pp. 1305-1314.

Bardzell J., Bardzell S., and Hansen L.K. (2015):
Immodest Proposals: Research Through Design and Knowledge[35],
Proc. CHI '15, ACM, New York, NY, USA, pp. 2093-2102.

Fox S. and Rosner D.K. (2016):
Continuing the Dialogue: Bringing Research Accounts Back into the Field[36],
Proc. CHI '16, ACM, New York, NY, USA, pp. 1426-1430.

Dalsgaard P. (2016):
Experimental Systems in Research through Design[37],
Proc. CHI '16, ACM, New York, NY, USA, pp. 4991-4996.

Information Systems / Software Design / Computer Sciences

Brooks F.P. (2010):
The Design of Design,

---

[33] http://dx.doi.org/10.1145/2702123.2702466
[34] http://dx.doi.org/10.1145/2702123.2702567
[35] http://dx.doi.org/10.1145/2702123.2702400
[36] https://doi.org/10.1145/2858036.2858054
[37] https://doi.org/10.1145/2858036.2858310

Addison-Wesley Professional.

Hevner A.R., March S.T., Park J., and Ram S. (2004):
Design science in information systems research,
MIS Q. 28(1), pp. 75-105.

Wieringa R. (2009):
Design science as nested problem solving[38],
In *Proc. DESRIST '09*. ACM, New York, NY, USA, Article 8.

desrist.org (2016):[39]
design science research in information systems and technology

Brooks F.P. (1995):
The Mythical Man-Month,
Addison-Wesley Professional; 2nd edition.

Glass R. L. (2006):
Software Creativity 2.0,
developer.* Books, 2006.

Knuth D.E. (1974):
Computer Programming as an Art – Turing Award Lecture[40],
Comm. ACM 17 (12), pp. 667–673.

Interview, Donald Knuth: A life's work interrupted[41],
Com. ACM 51, 8 (Aug. 2008), pp. 31-35.

Interview, Donald Knuth: The 'art' of being Donald Knuth[42],
Commun. ACM 51, 7 (July 2008), pp. 35-39.

Knuth D.

---
[38] http://wise.vub.ac.be/thesis_info/Design_Science_Wieringa.pdf
[39] desrist.org
[40] http://dx.doi.org/10.1145/361604.361612
[41] http://dx.doi.org/10.1145/1378704.1378715
[42] http://doi.acm.org/10.1145/1364782.1364794

The Art of Computer Programming (TAOCP)[43],
A comprehensive monograph that covers many kinds of programming algorithms and their analysis

Broy M. (2011):
Can Practitioners Neglect Theory and Theoreticians Neglect Practice?[44],
IEEE Computer 44(10), pp. 19-24.

Gamma E., Helm R., Johnson R., and Vlissides J. (1995):
Design Patterns: Elements of Reusable Object-Oriented Software,,
Addison-Wesley Longman Publishing Co., Inc.

Glass R. L., Ramesh V., and Vessey I. (2004):
An analysis of research in computing disciplines[45],
Commun. ACM 47(6), pp. 89-94.

Learning Sciences

Design-Based Research in Education[46] @ EduTech Wiki

Design-Based Research[47] @ Wikipedia

Edelson D.C. (2002),
Design Research: What We Learn When We Engage in Design,
Journal of the Learning Sciences 11(1), pp. 105—121.

Akker J. van den, Gravemeijer K., McKenny S., and Nieveen N. (Eds.) (2006):
Educational Design Research[48]

---

[43] http://www-cs-faculty.stanford.edu/~uno/taocp.html
[44] http://doi.ieeecomputersociety.org/10.1109/MC.2011.305
[45] http://dx.doi.org/10.1145/990680.990686
[46] http://edutechwiki.unige.ch/en/Design-based_research
[47] http://en.wikipedia.org/wiki/Design-based_research
[48] http://p4mristkipngawi.files.wordpress.com/2011/08/educational-design-research.pdf

Oxon: Routledge.

Barab S. A., & Kirshner D. (Eds.). (2001):
Rethinking methodology in the learning sciences, 44 *[Special Issue] J. Learning Sciences 10(1&2).*

Barab S. A., & Squire K. (Eds.). (2004):
Design-based research,
[Special Issue] Journal of Learning Sciences 13(1).

Brown A. L. (1992):
Design experiments: Theoretical and methodological challenges in creating complex interventions in classroom settings[49],
J. Learn. Sci. 2(2), pp. 141–178.

Cobb P., diSessa A., Lehrer R., and Schauble L. (2003):
Design experiments in educational research,
Ed. Researcher 32(1), pp. 9-13.

Collins A. (1990):
Toward a Design Science of Education,
New York: Bank Street College of Education.

Collins A., Joseph D., Bielaczyc K. (2004):
Design Research: Theoretical and Methodological Issues[50],
Journal Of The Learning Sciences 13(1), pp. 15-42.

Erickson F. & Gutierrez K. (2002):
Culture, rigor, and science in educational research[51],
Educational Researcher, 31(8), pp. 21-24.

Kelly A. E. (Ed.) (2003):
The role of design in educational research,
[Special Issue] Educational Researcher 32(1).

---

[49]http://www.cs.uml.edu/ecg/projects/cricketscience/pdf/brown-1992-design-experiments.pdf
[50]http://treeves.coe.uga.edu/EDIT9990/Collins2004.pdf
[51]http://www.jstor.org/stable/3594390

Maxwell J. A. (2004):
Causal explanation, qualitative research, and scientific inquiry in education[52]
Ed. Researcher 33(2).

Sandoval W.A., Bell P. (Eds.) (2004):
Design-based research methods for studying learning in context,
[Special Issue] Educational Psychologist 39(4).

---

[52]http://wwwedu.oulu.fi/tohtorikoulutus/jarjestettava_opetus/maxwell_scientific_inquiry.pdf

www.ingramcontent.com/pod-product-compliance
Lightning Source LLC
Chambersburg PA
CBHW071433180526
45170CB00001B/324